Letts
EDUCATIONAL

Exam Practice
A LEVEL

D0533004

A Level
Exam Practice

Covers AS and A2

Biology

Author

John Parker

Contents

AS and A2 exams

Different types of examination questions

Structured questions

Structured questions are in several parts. Each will normally begin with a brief amount of stimulus material, in the form of a diagram, data or graph. The parts are usually about a common context and they often progress in difficulty. They may start with simple recall, then test understanding of a familiar or unfamiliar situation.

The most difficult part of a structured question is usually at the end. Ascending in difficulty, a question allows a candidate to build in confidence. Right at the end technological and social applications of biological principles give a more demanding challenge. Most of the questions in this book are structured questions. This is the main type of question used in the assessment of both AS and A2 Biology.

Extended answers

In A2 and AS Biology questions requiring more extended answers usually form part of structured questions. These extended answers normally appear at the end of structured questions and typically have a value from four to twenty marks. These questions are allocated more lines, so you can use this as a guide as to how many points you need to make in your response. Typically for an answer worth ten marks the mark scheme would have around 12 → 14 creditable answers. You can score up to the maximum, ten marks. Extended answers are used to allocate marks for the **quality of communication**.

Candidates are assessed on their ability to use a **suitable style of writing, and organise relevant material, both logically and clearly**. The use of **specialist biological terms** in context is also assessed. **Spelling, punctuation and grammar** are also taken into consideration.

What examiners look for

Examiners are looking for logical answers to their questions so it is important to read the stimulus material carefully. Your answer will be credited if it contains the main facts. You do not get extra marks for writing a lot of additional words. Keep to the point! Make sure that your answer is clear, easy to read and concise. You need to apply your understanding of the concepts you have learned to unfamiliar situations given in the questions. You need to link together different topics. These are examined together in the synoptic papers, taken in the final stages of A2.

What makes an A, C and E grade candidate?

This is useful information if you wish to achieve your best potential grade. The way to do this is to make sure that you have a good, all round, knowledge and understanding of Biology.

A grade candidates recall and use biological knowledge, facts, principles and concepts from the complete specification, displaying only minor gaps in knowledge and understanding. Select biological knowledge for responses to questions, which is both relevant and logically presented. Use a range of specific technical terms is used in context. Carry out a range of calculations accurately, in a logical manner. The minimum mark for an A grade candidate is 80%.

C grade candidates recall and use biological knowledge, facts, principles and concepts reliably, from many parts of the specification; frequently select biological knowledge for responses to questions, which is in context and presented clearly and logically; frequently use appropriate technical terms; carry out a limited range of calculations accurately. The minimum mark for a C grade candidate is 60%.

E grade candidates demonstrate limited understanding of biological knowledge, facts, principles and concepts from some parts of the specification. Show understanding of basic principles and concepts beyond that expected of sound GCSE candidates. Use of some basic technical terms across the specification. Carry out a limited range of straightforward calculations. The minimum mark for an E grade candidate is 40%.

Successful revision

Revision skills

- Begin your revision programme several weeks before your examination

- Count-down to your exam day – small amounts of revision increasing in time as you approach the examination will be beneficial

- Writing down basic points on cards will help commit concepts to long term memory – it is important to have a reservoir of knowledge from which to retrieve facts during your exam.

- Last minute exam-cram is of little use. You need to assimilate the information gradually and apply the concepts to the new situations you will meet in the exam modules.

- Learn processes in sequence, carefully. Less able candidates miss out important parts of their answers and often corrupt the topics they have been taught. They are very few marks for being *almost correct!*

- Learning takes time. Success does not come easily. How hungry are you to earn your success?

- The most important part of your revision programme is to answer questions. The more you answer the better you will become.

- Without feedback, answering questions would be of no value. However, coupled with guidance from your teacher and this book, your strengths and weaknesses will be identified. Concentrate on weak area topics then see your performance improve … dramatically!

Practice questions

This book specialises in both questions and feedback. For the perfect solution use the Letts AS and A2 Guides to support you. Their bullet point style follows the way mark schemes are written. The main points regularly examined are included and published in an easy-to-learn format.

- Look at the answers of the A and C candidates. See if you could have done better.

- Check out the points missed by the C candidates as well as the major errors. This book includes errors regularly made by C, D and E candidates. Noting regular misconceptions will help you to avoid the same errors.

- Try to do the exam practice questions then check them out against the mark scheme.

- Make sure that you understand why the answers given are correct.

- Going through the chapters should help you prepare for specific modules.

- Communication marks are available for structured, logical answers with suitable spelling skills demonstrated.

- Finally, when you feel ready, it is time for the ultimate challenge. Try the synoptic paper!

Use the book regularly and use a highlighter pen to remind you of key information.

Planning and timing your answers

- Scan each question before you begin to answer. The key information is the stimulus material at the beginning.

- Try to link the question with a specific part of the syllabus. You will then be able to remember the principles which are being tested. Remember that examiners often repeat similar questions.

- Check out the mark allocation on the right of each question. This shows how many points you need to make in your answer.

- Plan your answers. Writing the first thing that comes into your head gains little credit. Keep to the point. Never give extra information just because you think you know it!

- Extended answers are usually at the back of a paper. Failure to reach the final question can penalise you much more than missing the first question. Many candidates answer the last question first.

- Present you answers clearly. There is nothing to gain by hurrying your responses so much that the examiner cannot understand your writing. Aim for clarity.

How to boost your grade

Learn all definitions throughout the specification. These are straightforward marks to gain, but you must prepare well. Along with labelling diagrams, they are among the easier marks to answer successfully.

Graphs appear in a lot of questions. Be ready to interpret the relationship shown by the data. If asked to **describe** the relationship shown, this is simply a comment linking the two variables, *e.g. the amount of product produced in an enzyme catalysed reaction increases to a peak, then falls with temperature increase.*

If the question requires you to **explain** the relationship then the reasons for the changes in data are required. In this example you would need to refer to the shape of the active site of the enzyme molecules and collisions between substrate and enzymes at different temperatures.

In explaining relationships, when more than one graph line is shown, look for peaks and troughs. Where a peak of one graph line corresponds with a trough of another then it is likely that they are linked, e.g. where pressure in a heart atrium is high, and the volume is reducing then a student could conclude that the atrium was both contracting and emptying.

Where a change takes place, then you may be expected to think of a reason not shown on the graph, *e.g. the number aphids increases dramatically from April to September. You may be expected to answer in terms of higher Spring temperatures. Greater plant growth in Spring should be related to additional food for the aphids so reproduction takes place.*

Detail is required at this level. The following example will illustrate this point. A question requires the candidate to explain how oxygen is passed to the tissues.

The candidate writes, "haemoglobin picks up oxygen and transports it to the tissues. Here the oxygen passes from a red blood cell to the cells."

This answer lacks the required detail! This is the detail required.

The red blood cells contain haemoglobin ;
Haemoglobin has an affinity for oxygen so it binds to the haemoglobin;
At the tissues carbon dioxide is formed during respiration;
The carbon dioxide reduces the affinity of haemoglobin for oxygen;
so some oxygen is off-loaded at the tissues;
this is known as Bohr shift;

The first answer shows a candidate still displaying GCSE ability. AS and A2 demand much more detail. Set out to learn all the details. Learn it all in logical steps.

This book shows the detail of official mark schemes. Go for ultimate detail and achieve the ultimate grade, A!

You need to do **Regular Revision** through the course; this keeps the concepts "hot" in your memory, "simmering and distilling", ready to be **retrieved** and **applied** in the synoptic contexts.

Questions with model answers

C grade candidate – mark scored 4/6

 For help see Revise AS Study Guide pages 24 and 25

(1) The diagram below shows the structure of a lipid molecule.

```
        H    O
        |    ||
    H—C—O—C\/\/\/\/\/\/\/\/\/\
        |    O
        |    ||
    H—C—O—C\/\/\/\/\/\/\/\/\/\
        |    O
        |    ||
    H—C—O—C\/\/\/\/\/\/\/\/\/\
        |
        H
        |_____||_____|
            A              B
```

Examiner's Commentary

(a) Name the parts labelled A and B.   **[2]**

   A Glycerol ✔

   B Fatty acid ✔

That was a good start! The candidate was able to recognise the components of a simple lipid.

(b) Name this type of lipid.   **[1]**

   Glyceride ✗

Close but not quite good enough! The answer required by the Exam Board was triglyceride. Note the three fatty acids bonded to the glycerol.

(c) Name the chemical reaction used to form bonds between A and B.   **[1]**

   Condensation ✔

Yes! Correct. As the glycerol and fatty acid bond together water is given off. This is condensation.

(d) State ONE function of this type of lipid in living organisms.   **[1]**

   The molecule is an energy source ✔.

Correct. The molecule is also involved in waterproofing and is an energy store. These responses would also have been credited.

(e) State ONE feature of the molecules of this type of lipid that makes them suitable for the function you have given.   **[1]**

   The molecules are insoluble ✗.

[Edexcel Specimen]

Not correct. The molecule is insoluble but this property is not associated with the 'energy source' given in response to (d). Take care with linked marks! It would have been correct if the answer to (d) had been 'waterproofing'.

Questions with model answers

A grade candidate – mark scored 14/15

For help see Revise AS Study Guide pages 24 and 25

(2) PrP is a protein normally found in brain tissue. The diagram shows part of the structure of a PrP molecule. The dotted line represents the middle part of the molecule which has been left out of this diagram.

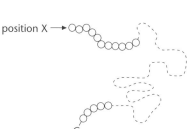

Examiner's Commentary

(a) **(i)** Name the monomers which make up this molecule. **[1]**

Amino acids ✔

The key to scoring this mark is the – COOH (carboxylic acid group). A polymer with this group at one end points to a polypeptide. Polypeptides consist of linked amino acids.

(ii) Name ONE chemical element which would be found in this molecule but not in a polysaccharide molecule. **[1]**

Nitrogen ✔

Correct! The exam board accepted sulphur from the sulphur bridges, although none are shown in the diagram.

(iii) Give the formula of the chemical group that would be found at position X on the molecule. **[1]**

Amine group ✗

*No! That was a slip up. The candidate probably knew the answer but failed to give the **formula**, NH_2. Even high ability candidates slip up sometimes but an A is still possible! You don't have to score 100% to be awarded an A.*

PrP molecules are found on the outside of the cell surface membranes of nerve cells. The precise function of the PrP is still unknown but it is thought that its tertiary structure enables it to act as a receptor molecule.

(b) Describe:

(i) the secondary structure of a protein **[1]**

α helix or β pleated sheet ✔

Correct. The candidate gave two correct answers, where only one would have been enough. Valuable time was lost.

(ii) the tertiary structure of a protein. **[1]**

Further folding of a polypeptide ✔.

Correct. A globular shape would also have been credited, this being the result of the further folding already given by the candidate.

A grade candidate continued

 For help see Revise AS Study Guide pages 24 and 25

(c) **(i)** What is meant by a receptor molecule? **[1]**

Other molecules, e.g. a specific hormone, may fit into a specific receptor molecule ✔.

(ii) Explain how its tertiary structure might allow a protein molecule to act as a receptor molecule. **[2]**

Tertiary structure gives the receptor molecule its precise shape ✔.

It provides a site into which another molecule may fit ✔.

(d) It appears that when a cow gets BSE something causes the PrP molecules to become sticky so that they clump together. With an electron microscope, string-like fibrils composed of clumps of PrP can be seen in the brains of cattle affected by BSE. Explain why the string-like fibrils of PrP can be seen in the brain tissue of cattle with an electron microscope but not with a light microscope. **[2]**

The resolution of the light microscope is not as good as that of the electron microscope ✔.

The wavelength of the electron beam is smaller ✔.

(e) It is not known what makes the PrP molecules stick together. One hypothesis is that an unknown infectious agent may bring about a change in the secondary structure of the PrP molecule. This could explain the fact that the PrP molecules in fibrils are resistant to the action of protein digesting enzymes which bring about the hydrolysis of PrP from healthy animals.

(i) Describe what happens during hydrolysis of PrP. **[2]**

PrP breaks down into amino acids ✔.

Water is used in the reaction ✔.

(ii) Suggest how a change in the secondary structure of the PrP molecule could explain the fact that the PrP in fibres is resistant to the action of protein digesting enzymes. **[3]**

Change in the secondary structure changes its shape ✔.

The new shape does not fit ✔ into the active site of the enzyme ✔.

[AQA B Specimen]

Examiner's Commentary

*This is a strong response. The Exam Board would have accepted the idea that a molecule may fit into a receptor. This candidate shows 'A' ability because of the knowledge of **specificity**. Not just any molecule will fit into a receptor molecule. Its size and shape must be precise.*

*Full marks again! Take care with 'receptor' questions. The receptor has a site into which a molecule fits but this is **not** an active site. Reserve this term just for enzymes.*

*Full marks. The candidate suggested that the resolution of the electron microscope was greater than that of the light microscope, **by implication**. It would have been better to directly state the greater resolution of the electron microscope.*

*Full credit given. The peptide bonds are broken down to release the amino acids. **All** hydrolysis reactions need the addition of water.*

All correct! The candidate demonstrated understanding of both protein structure and enzyme action.

Exam practice questions

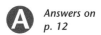

Answers on p. 12

(1) The statements in the table refer to three polysaccharide molecules. If the statement is correct place a tick (✔) in the appropriate box and if the statement is incorrect place a cross (✗) in the appropriate box.

Statement	Starch	Glycogen	Cellulose
Polymer of α glucose			
Glycosidic bonds present			
Unbranched chains only			
Energy store in animal cells			

[Edexcel Specimen]

(2) The diagram shows how some biological molecules may be separated from each other ...

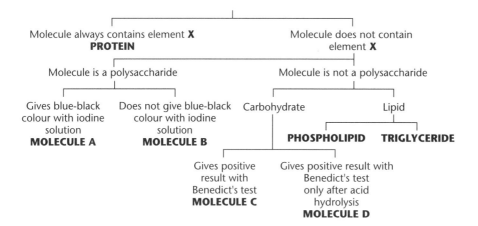

(a) Name element X. **[1]**

(b) Which of molecules A to D could be:
  (i) glucose **[1]**
  (ii) glycogen **[1]**

(c) Describe a chemical test which could be used to confirm the presence of a protein. **[2]**

(d) Describe **one** way in which the structure of a phospholipid differs from that of a triglyceride. **[2]**

[AQA A Specimen]

(3) (a) For each of the following pairs of carbohydrates, give one word, which describes **both** of the molecules and **distinguishes** them from the other pairs.
  (i) Cellulose and glycogen   **(iii)** Ribose and deoxyribose
  (ii) Maltose and lactose   **(iv)** Glucose and fructose **[4]**

(b) Name the reagent for testing for reducing sugars in food. **[1]**

(c) Name a sugar that would not give a positive test with this reagent. **[1]**

[WJEC Specimen]

(4) A dipeptide was heated with an acid for 20 minutes. Three different samples were taken and loaded onto a piece of chromatography paper. These were:
1. the dipeptide only
2. the dipeptide after it had been heated with the acid for 2 minutes
3. the dipeptide after it had been heated with the acid for 20 minutes.
The resulting chromatogram is shown in the drawing.

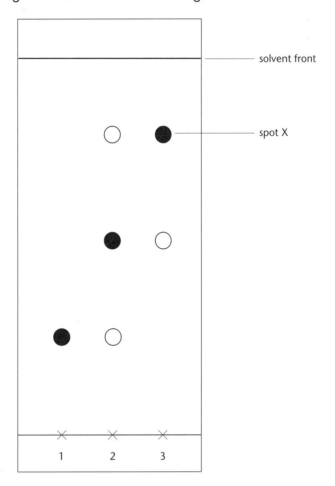

(a) What type of reaction results in the conversion of a dipeptide to amino acids? **[1]**

(b) Explain why three spots were obtained when the dipeptide was heated with the acid for two minutes. **[2]**

(c) **(i)** Calculate the R_f value of spot X. Show your working. **[2]**
(ii) Explain how you could use the R_f values to identify spot X. **[1]**

[AQA A Specimen]

Answers

(1)

Statement	Starch	Glycogen	Cellulose
Polymer of α glucose	✔	✔	✗
Glycosidic bonds present	✔	✔	✔
Unbranched chains only	✗	✗	✔
Energy store in animal cells	✗	✔	✗

Examiner's tip

Remember that glycosidic bonds link the monomers in all polysaccharides. Both starch and glycogen have branched chains.

(2) (a) Nitrogen   **(b) (i)** Molecule C   **(ii)** Molecule B

Examiner's tip

This question is based on a dichotomous key where you have to make a number of key decisions. Glucose is the only reducing sugar in the diagram and glycogen has no reaction with iodine.

(c) Biuret reagent(s)/sodium hydroxide and copper sulphate; lilac/mauve/purple;

(d) Phospholipid has a phosphate group; This phosphate replaces a fatty acid molecule in a triglyceride.

Examiner's tip

Remember that these molecules are very similar. The only difference is that a triglyceride has three fatty acid residues and a phospholipid only two. The missing fatty acid is replaced by a phosphate group.

(3) (a) (i) Polysaccharides   **(ii)** Disaccharides

(iii) Pentoses   **(iv)** Hexoses

Examiner's tip

Cellulose and glycogen have many monomers and are polysaccharides.
Maltose and lactose are both disaccharides with 2 monomer units.
Ribose and deoxyribose are both 5C pentoses. Glucose and fructose are both 6C hexoses.

(b) (i) Benedict's reagent/Fehlings reagent   **(ii)** Sucrose

(4) (a) Hydrolysis   **(b)** Some of the dipeptide/2 different amino acids

Examiner's tip

Some of the dipeptide was unchanged, hence the spot near the start line. The peptide bond was broken by hydrolysis, into two different amino acids.

(c) (i) 8 cm/10 cm = 0.8

Examiner's tip

You should measure the distance moved by the spot on the diagram, then measure the distance moved by the solvent front. R_f is distance moved by the spot divided by the solvent front i.e. 8/10 = 0.8

(ii) Check with table of known values or compare with R_f value of known substance.

Examiner's tip

R_f value is always the same for the same substance in same solvent under same conditions

Questions with model answers

C grade candidate – mark scored 4/6

 For help see Revise AS Study Guide pages 33 and 35

Examiner's Commentary

(1) The diagram below shows the structure of a liver cell as seen using an electron microscope.

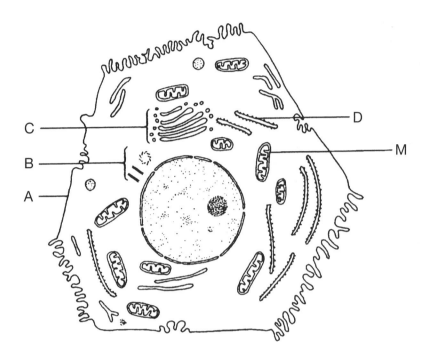

(a) Name the parts labelled A, B, C, and D. **[4]**

A Plasma membrane ✔   B Centriole ✔

C Golgi apparatus ✔   D Endoplasmic reticulum ✗ ←

A, B and C had a number of other terms which could be credited. The candidate did not give enough detail for D. Rough endoplasmic reticulum was correct or ribosome. Endoplasmic reticulum can be <u>rough</u> or <u>smooth</u>. Just giving endoplasmic reticulum would not be enough for credit.

(b) The magnification of the diagram is × 12 000. Calculate the actual length of the mitochondrion labelled M, giving your answer in μm. Show your working. **[2]**

Length of mitochondrion diagram · 12 mm

True length = 12/12 000 = 0.001 mm ✔ ←

length = 0.001 mm ✗

[Edexcel Specimen]

*The candidate measured the diagram correctly. The true length is 12 000 times smaller. However, the answer was required in μm.
There are 1000 μm in one mm
Length 0.001 × 1000 = 1.0 μm   (Tip: always check your units!)*

13

Questions with model answers

A grade candidate – marks scored 12/12

 For help see Revise AS Study Guide pages 33–35 and 39–40

Examiner's Commentary

2. The photograph is an electronmicrograph of a mammalian pancreas cell (× 15 000)

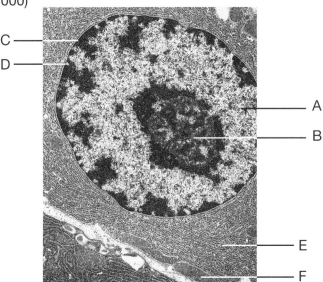

(a) Identify the structures labelled A–F. **[6]**

A Nucleus ✔   B Nucleolus ✔

C Nuclear envelope ✔   D Pore in nuclear envelope ✔
(Nuclear membrane is also correct.) (Nuclear pore is also correct.)

E Rough endoplasmic reticulum ✔   F Mitochondrion ✔

Photographs can be difficult to analyse. Here the 'A' candidate realised that most of the photograph is taken up by the nucleus. After this identification became less difficult. The first task is always to deduce how many cells there are.

(b) (i) Explain why region B is more darkly stained than region A. **[2]**

Region B has incorporated a lot of heavy metal stain, e.g. osmium salts ✔. Very few electrons can pass through this region ✔.

Good answer. Many electrons are blocked by the stained part. Unable to reach the screen an electron shadow appears on screen.

(ii) State one disadvantage of electron microscopy. **[1]**

The specimens are not alive, so may have changed in some way ✔.

*Additionally there is a greater chance of **artefacts** being part of the image.*

(c) Calculate the actual diameter of the organelle in micrometres (µm) between the two parts X on the electronmicrograph. Show your calculations in the space provided. **[2]**

Length between the two points was 162 mm ✔
Answer = 162/15 000 mm × 1000 = 10.8 µm ✔

The candidate remembered all the stages, measured accurately and remembered that there are 1000 µm in 1 mm.

(d) Give **one** piece of evidence, visible in the electronmicrograph, which suggests that the main function of this cell is protein synthesis. **[1]**

There is a lot of rough endoplasmic reticulum with clearly visible ribosomes ✔.

Another strong answer. Ribosomes have a major role in the synthesis of proteins.

[NICCEA Specimen]

Exam practice questions

A *Answers on p. 16*

(1) Some cells were broken up and the organelles they contained were separated by ultracentrifugation. The drawing shows three types of organelle which were obtained.

organelle X   organelle Y   organelle Z

(a) The cells were all the same type. Which of the cells A to D listed below might they have been?

    A bacterial cells
    B red blood cells
    C cells from a plant leaf
    D epithelial cells from the lung   **[1]**

(b) Explain why only organelle X appeared in the sediment when the broken up cells were centrifuged at the lowest speed.   **[1]**

(c) Give the function of:

    (i) organelle Y
    (ii) organelle Z.   **[2]**

(d) Explain why an electron microscope would be required to see the detailed structure of organelle Y.   **[2]**

[AQA A Specimen]

(2) (a) The following table lists some of the features of cells. Complete the table by ticking (✔) in the appropriate column(s) if the feature listed is found in eukaryotes, prokaryotes or both cells.   **[8]**

Feature	Eukaryotes	Prokaryotes
Usually less than 10 μm in size	✔	✔
Mitochondria present	✔	✔
Respiratory enzymes present	✔	
Ribosomes present	✔	
DNA usually a continuous loop		✔
Presence of nuclear membrane	✔	

(b) Name one way in which the cell wall of most prokaryotic cells differs from that of a plant cell.   **[1]**

(c) **(i)** What class of chemical usually forms the outer coat of a virus particle?   **[1]**

    (ii) What is the coat called?   **[1]**

[WJEC Specimen]

Answers

(1) (a)   C

(b)   Chloroplasts are the most dense of these organelles and sink first.

Examiner's tip

In this part of the question you need to be aware of the stages of ultracentrifugation. At the slowest speed the nuclei are the most dense and are in the lowest sediment. An assumption you need to make here is that the nuclei must have been removed earlier.

(c) (i) Aerobic respiration
(ii) Protein synthesis

Examiner's tip

*Learn all of the functions of the organelles. The Exam Board credited respiration but could have insisted on aerobic for part **(i)**. Ribosomes on the rough endoplasmic reticulum clearly point towards protein synthesis.*

(d)   Needs higher resolution/must distinguish between structures very close together; uses a shorter wavelength (electron beam).

(2) (a)   Mark each line ticked as follows:
Line 1 P
Line 2 E
Line 3 E, P
Line 4 E, P
Line 5 P
Line 6 E

Examiner's tip

This type of question requires you to make a number of decisions. Remember key facts about both eukaryotic and prokaryotic organisms. Prokaryotes do not have membrane bound organelles, so they have no true nucleus or mitochondria.

(b)   Plant cell wall is cellulose (bacteria always other chemicals)

(c) (i) Protein
(ii) Capsid

Examiner's tip

Outer coat of viruses is made of a number of capsomeres each of which is a protein unit.

Questions with model answers

C grade candidate – mark scored 8/11

 For help see Revise AS Study Guide pages 46–48

(1) (a) Define the term 'enzyme'. **[2]**

> *An enzyme is a molecule which catalyses reactions in living organisms* ✔ *by lowering the activation energy* ✔

Examiner's Commentary

Activation energy is the amount of energy needed to make the reaction begin.

(b) Explain why enzymes are essential for metabolism. **[2]**

> *Lowering the activation energy allows reactions to take place at the usual physiological temperature of the organism, such as 37 °C in humans* ✔*.*

The second part of the answer was missed. Enzymes allow reactions to take place 'at physiological pH'.

(c) The following diagram shows the influence of substrate concentration on the rate of an enzyme-controlled reaction working at its **optimum temperature**.

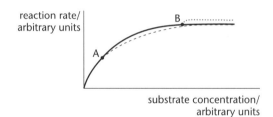

Explain fully the shape of the curve. **[2]**

> *The rate of reaction increases as the substrate increases* ✗*.*
>
> *Finally the maximum rate of reaction is reached when all of the active sites are occupied* ✔*.*

*The first part was a typical error. It was a **description**, and did not **explain**. The candidate should have referred to the fact that active sites were available at lower substrate concentration so any increase in substrate concentration increases the rate.*

(d) (i) If the temperature was slightly increased at point A, draw on the graph a line to show how the plotted curve would change. *See the line (- - - - - - -) on the graph* ✔*.* **[1]**

The line is correct.

(ii) Explain this change.

> *The enzyme was working at optimum temperature* ✔*, a higher temperature partially denatures the enzyme so the rate is lower* ✔*.* **[2]**

Given a reaction at optimum temperature, at a higher or lower temperature the rate, with concentration would rise more slowly before levelling off.

(e) (i) If a small additional amount of enzyme was added to the reaction at point B, draw on the graph a line to show how the plotted curve would change.

> *See the line (·············) on the graph* ✔*.* **[1]**

(ii) Explain this change. **[1]**

> *More enzyme molecules available for substrate molecules* ✗*.*

*More active sites are available, so more substrate molecules can **bind** with them so the rate increases.*

[WJEC Specimen]

Questions with model answers

A grade candidate – mark scored 9/10

For help see Revise AS Study Guide page 53

Examiner's Commentary

(2) **(a)** The diagrams below represent two alternative commercial production systems involving enzymes. In the batch reactor, a fixed amount of soluble enzyme and substrate are mixed together in a solution. In the continuous-flow column reactor, substrate molecules flow past enzymes which are immobilised on an inert support material.

Batch Reactor

stirrer

enzyme + substrate

enzyme/product mixture

purification

enzyme   product

Column Reactor

substrate

immobilised enzyme

inert support material

product

Identify **two** advantages that the immobilised enzyme system has over the dissolved enzyme system. Explain your answer. **[4]**

Immobilised enzymes adhere to the inert support material, so can be used continuously ✔*, because the enzymes remain, this is more economic* ✔*. The enzymes are withheld in the reactor and so do not contaminate the product* ✔ *so no purification is needed* ✔*.*

Often, immobilised enzymes work just as well as freely moving ones. Here the graph suggests that some of the active sites of some enzyme molecules have been affected.

(b) One disadvantage of enzyme immobilisation is that it may change the catalytic activity of the enzyme as shown in the graph below.

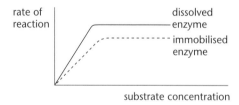

rate of reaction

dissolved enzyme

immobilised enzyme

substrate concentration

A grade candidate continued

For help see Revise AS Study Guide page 53

Examiner's Commentary

Suggest **two** reasons for the observed change in enzyme activity when immobilised. **[4]**

> *The active sites of some enzyme molecules could be covered by an 'adhesive' material ✔, so that the binding of some substrate molecules could be prevented ✔. As the substrate molecules flow past the enzymes there is less of a chance for the substrate molecules to bind with the active sites ✔, so there is a slower rate of enzyme-substrate complex ✔.*

Full marks! As the substrate molecules flow along there is a tendency for the rate to be reduced if it is more difficult to contact an active site.

(c) Immobilisation is generally accepted as a strategy to develop more stable enzyme preparations. Thus immobilisation often increases the temperature and/or pH range over which an enzyme remains active.

Suggest how immobilisation may improve enzyme stability. Explain your answer. **[2]**

> *There is strong bonding between the enzyme and the inert support material ✔ the enzyme does not change ✗.*

First part correct. Second part not correct. Immobilisation helps withstand the disruption of the tertiary structure of the enzyme by higher temperature or pH change.

[NICCEA Specimen]

Exam practice questions

 *Answers on p. 21*

(1) Catalase is an enzyme that breaks down hydrogen peroxide into oxygen and water. The activity of catalase can be measured by soaking small discs of filter paper in a solution containing the enzyme. The discs are immediately submerged in a dilute solution of hydrogen peroxide. The filter paper discs sink at first but float to the surface as oxygen bubbles are produced. The reciprocal of the time taken for the discs to rise to the surface indicates the rate of reaction.

An experiment was carried out to investigate the effect of substrate concentration on the activity of catalase. A filter paper disc was soaked in a solution containing catalase, and then submerged in a buffer solution containing hydrogen peroxide. The time taken for the disc to rise to the surface was recorded. The experiment was repeated using a range of concentrations of hydrogen peroxide.

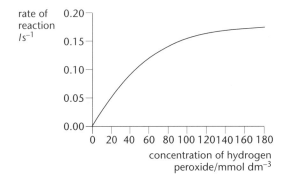

(a) State why a buffer solution was used in this experiment. **[1]**

(b) Describe the relationship between the rate of reaction and the concentration of hydrogen peroxide. **[3]**

(c) Explain this relationship between substrate concentration and the rate of reaction. **[4]**

(d) Describe how a solution containing 160 mmol of hydrogen peroxide per dm^3 would be diluted to prepare a solution containing 80 mmol of hydrogen peroxide per dm^3. **[2]**

(e) Describe how this experiment could be modified to investigate the effect of temperature on the activity of catalase. **[2]**

(2) (a) Complete the series of diagrams below to show how a catabolic reaction would take place, in terms of the induced fit theory. Label your drawings. **[3]**

(b) Explain TWO ways that an enzyme catalysed reaction can be prevented by the presence of molecules other than the substrate. **[3]**

Answers

(1) (a) To keep pH constant/enzymes are affected by pH.

Examiner's tip

Always be prepared to use the term 'buffer'. It keeps the pH constant even though more H^+ ions would increase acidity or more OH^- ions would increase alkalinity.

(b) Rate increases as the concentration of substrate increases.
There is a constant increase (approximately) between 20–80 mmol dm^3.
Rate levels off after this.

(c) (In terms of kinetic theory) number of collisions increase with the increase of substrate concentration, as a result the rate increases, until all of the active sites are in use, rate is then constant or maximum.

Examiner's tip

*Collision theory is about the **random** impacts between the active site of the enzymes and substrate molecules. As the number of substrate molecules increases the **chance** of suitable collisions increases.*

(d) Mix equal volumes of hydrogen peroxide solution and distilled water or deionised water or buffer solution.

(e) Suggest a range of temperatures (at least three), use of same substrate concentration, use the same volume of hydrogen concentration (standardise pH), allow the substrate to equilibrate before adding the filter paper disc, discs must be uniform in size,
use of same enzyme concentration, repeat at each temperature, plot a graph of the rate of reaction against temperature.

Examiner's tip

This tests a key skill, your ability to design or modify an investigation.
*Note that there is usually a mark for **replication**. Here a range of at least three temperatures must be suggested, to earn a mark. Then another mark is available for repeating each chosen temperature. Remember this for your live examination!*

(2) (a) Draw the enzyme with active site changed to suit substrate shape, label the enzyme substrate complex, two or more products formed.

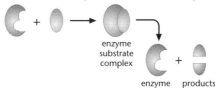

(b) Competitive inhibitor, molecules fit into active site preventing substrate entry.
Non-competitive inhibitor, molecules fit into alternative site, change shape of active site preventing substrate entry.
H^+ or OH^- ions or other correct named substance, which can denature enzyme changing active site.

Examiner's tip

Always remember that an inhibitor can prevent the reaction or reduce the rate of reaction.
A non-competitive inhibitor binds to a different part of the enzyme than the active site but it still changes the shape of the active site.

Questions with model answers

C grade candidate – mark scored 5/7

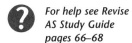

For help see Revise AS Study Guide pages 66–68

(1) The diagram below shows a human heart at a specific stage in the cardiac cycle.

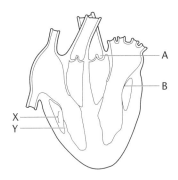

Examiner's Commentary

(a) Name the parts labelled A and B.   **[2]**

   A Semi-lunar valve ✔ *B Mitral valve* ✔

Both answers are correct. A is also called a pocket valve but was not in the mark scheme of this Exam Board. B is also called the left atrioventricular valve or the bicuspid valve.

(b) Name the stage of the cardiac cycle shown in the diagram and give TWO reasons for your choice.   **[3]**

   Name of stage: systole ✗

*The response is not accurate enough. **Atrial** systole would have been credited because the atria are contracting.*

   Reason 1: the mitral valve is open ✔.

   Reason 2: the semi-lunar valve is closed ✔.

Both of the reasons are correct and can be deduced from the diagram.

(c) Give ONE function of each of the parts X and Y.   **[2]**

   X prevents the valve being inverted, back into the atrium ✔.

   Y is an anchor point for the tendonous cords ✗.

[Edexcel Specimen]

The function of X is correct. The tendenous cords prevent the valve from collapsing back into the atrium. The response to the function of Y gains no credit. It adjusts the tension in the valve and actually contracts to cause this.

A grade candidate – mark scored 14/15

 For help see Revise AS Study Guide pages 61 and 62

(Quality of written communication will be assessed in this question)

(2) **(a)** Mesophyll cells in a plant use carbon dioxide for photosynthesis.

Describe how carbon dioxide from the atmosphere reached these mesophyll cells. **[2]**

Carbon dioxide enters the leaf by diffusion ✔, through the stomata and the air spaces ✔.

(b) The diagram below shows the way in which water flows over the gills of a fish. The graph shows the changes in pressure in the mouth cavity and in the opercular cavity during a ventilation cycle.

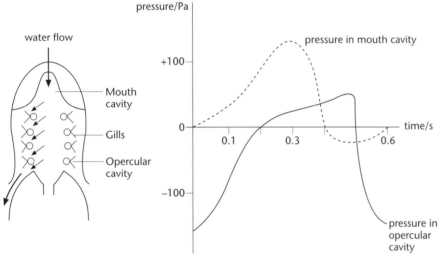

(i) Use the graph to calculate the rate of ventilation in cycles per minute. **[1]**

$$\frac{60\ s}{0.6} = 100\ cycles\ per\ minute\ ✔$$

(ii) For most of this ventilation cycle, water will be flowing in one direction over the gills. Explain the evidence from the graph which supports this. **[2]**

Pressure in the mouth cavity is higher than the pressure in the opercular cavity for the greater part of the cycle ✔, water flows in the direction from higher to lower pressure ✔.

Examiner's Commentary

*Correct. Two marks are awarded. The key part of the task is that the carbon dioxide must reach the mesophyll cells. Just entering via the stomata would not allow the CO_2 to reach the cells. **Both** 'entry via stomata' and 'through the intercellular air spaces' are needed.*

Note that a single cycle takes 0.6 seconds so the rate is calculated by the number of times a single cycle is divided into one minute. Always look for a repeated pattern in this type of question.

In any graph based question like this note that movement of fluids or gases is from higher to lower pressure. Here you needed to relate events on the graph to events inside the fish.

Questions with model answers

A grade candidate continued

 For help see Revise AS Study Guide pages 61 and 62

(iii) Explain how the fish increases pressure in the buccal cavity. **[2]**

The mouth closes ✔.

Muscles at the base of the mouth contract to allow the base to move upwards ✔.

(c) Explain how the gills of a fish are adapted to form a specialised exchange surface. **[8]**

The gill filaments are very thin ✔, *and there are many of them* ✔.

Each filament is lined with flattened epithelial cells ✔, *which gives a short diffusion path* ✔.

Water moves in the opposite direction to blood ✔, *which maximises the diffusion gradient along each filament* ✔.

The ventilation movements of the buccal cavity and opening and closing of the operculum continually allows water to cross the gills ✔.

[AQA B Specimen]

Examiner's Commentary

Remember to work out the answer if you do not know it directly. Pressure could only build up if the mouth is shut.

The candidate answered well but just fell short of the 8 marks available. The candidate was aware of the replacement of water but failed to describe the replacement of blood to the gills by the circulatory system. Deoxygenated blood returning to the gills allows a maximum rate of gaseous exchange.

Exam practice questions

A *Answers on p. 26*

(1) **(a)** **(i)** Define the term water potential [1]

(ii) What is the water potential of pure water? [1]

(b) The graph shows the change in cell volume and water potential values of a flaccid plant cell placed in distilled water.

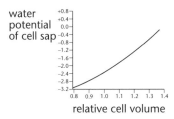

Using the graph,

(i) Describe the change in water potential of the cell. [1]

(ii) Explain your answer. [1]

(c) The **solute** potential of the cell is never zero. Explain. [1]

[WJEC Specimen]

(2) The mean concentration of sugar in the phloem sap of the stems of cotton plants was measured at four different times of the day and the results are plotted in the graph **(C)** below.

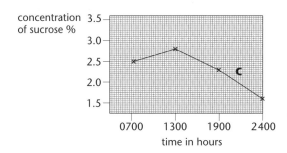

At the start of the experiment, a second batch of plants had a ring of tissue removed from the stem, destroying the phloem at the point where the ring was taken. The table shows the sugar concentration above the ring **(A)** and the concentrations below the ring **(B)**, at the same time intervals.

	Time / hours			
	0700	**1300**	**1900**	**2400**
A (above)	2.5	3.5	3.1	2.7
B (below)	2.5	1.9	1.6	1.5

(a) Name the conducting vessels in the phloem that were destroyed by the ringing process. [1]

(b) Plot the values for **(A)** and **(B)** on the graph above. [2]

(c) For the period 0700 hours to 1300 hours, explain **fully**:

(i) the difference between **(A)** and **(C)** [2]

(ii) the difference between **(B)** and **(C)**. [2]

(d) What is the reason for the trend shown by the plots between 1900 hours and 2400 hours? [1]

[WJEC Specimen]

Answers

(1) (a) (i) Water potential is the tendency of water to leave the cell.
     (ii) Zero

Examiner's tip

There must be a solute, such as mineral salts, in the water if water potential is to have value. This being so, it would be negative and the usual units would be pascals.

(b) (i) As the water potential of the cell becomes less negative the cell volume increases.
     (ii) Water enters the sap vacuole by osmosis, because the cell sap is hypertonic to the distilled water, as this happens the water potential changes.

Examiner's tip

*Always remember that water moves from areas of a **less negative water potential to a more negative water potential**. As water enters a cell the water potential value becomes less negative.*

(c) Cell sap always contains some mineral salts.

(2) (a) Sieve tubes

(b) One mark is taken off for each plotting error.

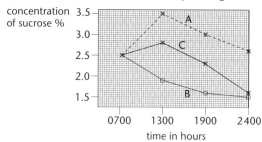

(c) (i) The sugar accumulates above the cut/ringing does not allow the sugar to pass down (the sugar is not being used quickly so production of sugar, by photosynthesis, exceeds the amount being used in respiration).

     (ii) The supply of sugar has been cut off by the ringing technique, and sugar is being used in respiration so that sugar concentration falls.

Examiner's tip

Ringing destroys the continuity of the phloem elements. The leaves above the damage continue to photosynthesise and so sugar production accumulates over the ringed part. Below, the roots have a big energy requirement, so sugar concentration drops quickly.

(d) Photosynthesis gradually stops, so the sugar concentration falls as the sugar is utilised in respiration.

Questions with model answers

C grade candidate – mark scored 9/15

For help see Revise AS Study Guide pages 82–85

Examiner's Commentary

(1) The diagram below shows part of a DNA molecule.

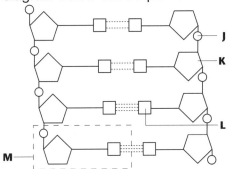

(a) (i) Name J to M **[4]**

J Phosphate ✔ K Pentose sugar ✔
L Organic base ✗ M Nucleotide ✔

*Almost full marks here. L is not accurate enough. The Exam Board required knowledge that the base is **nitrogenous**.*

(ii) What do the dotted lines in the diagram represent? **[1]**

Hydrogen bonds ✔

The bases always bind the two strands together by hydrogen bonds.

(b) State **three** ways in which the structure of messenger RNA differs from DNA. **[3]**

1 RNA is single stranded ✗.
2 RNA has uracil instead of thymine ✔.
3 The DNA and RNA are different sized strands ✗.

*Only one correct here. RNA **is** single stranded but the candidate had to clearly state that the DNA was double stranded. Both facts were required. RNA is much shorter than DNA. The answer given was ambiguous. Be perfectly clear when giving your answers.*

(c) Explain why exact replication of DNA is necessary. **[2]**

To reproduce genetic information ✗.
Because the same DNA will code for the same proteins ✔.

To reproduce genetic information was almost correct. If the word identical had been given to describe identical genetic information is produced then the mark could be awarded.

Part of a DNA molecule is shown below.

```
 ┌ AT ┐
 ├ TA ┤
 ├ GC ┤
 ├ CG ┤
 ├ TA ┤
 ├ TA ┤
 └ CG ┘
```

(d) In the space provided, show by means of a diagram what happens to this part of DNA during replication: **[4]**

Diagram shows:
the two strands, attached at the middle, begin to separate ✔, breaking at the hydrogen bonds ✔, and the bases of each strand binding with a new, complementary base, e.g. thymine with adenine ✔. .... ✗

The candidate only gave 3 parts to the answer and there were 4 marks. Always look carefully at the mark allocation! 'New nucleotides bind with each strand' would have been credited for the final mark.

(e) Name the enzyme involved in replicating the DNA molecule. **[1]**

DNA polymerase ✔.

[OCR Specimen]

Questions with model answers

A grade candidate – mark scored 8/8

 For help see Revise AS Study Guide pages 87–114

(2) Biologists in Australia are using genetic engineering to produce a gene to insert into oranges. They have combined two pieces of DNA to produce a gene which they have called SDLS-2.

The diagram shows part of this gene.

DNA sequence that switches on a gene used in seed formation   DNA sequence that kills plant cells

They intend to introduce this gene into orange trees.

(a) Describe how each of the following might be useful in this process. **[1]**

  (i) Ligase enzymes

  The ligase enzymes join the DNA pieces together ✔.

  (ii) A vector **[2]**

  The new gene is inserted into a vector first ✔,

  which is then taken up by a plant cell ✔.

(b) (i) Explain how this gene could lead to the production of seedless fruit. **[2]**

  The gene is switched on as seeds are formed ✔,

  so that the seed is finally killed during formation ✔.

  (ii) Describe the possible dangers which might result from growing orange trees containing the SDLS-2 gene. **[3]**

  The plants no longer have seeds, but they are needed for reproduction ✔.

  If the genes were able to become incorporated into other species, then they would become sterile ✔.

  If the gene which kills cells was switched on in other parts of a plant then it is likely that the complete plant would be killed ✔.

[AQA Specimen]

Examiner's Commentary

In genetic engineering there are a limited number of enzymes to remember. Note that ligase enzymes attach compatible ends of DNA together.

Full marks. Additionally the fact that the fruit cells are still able to develop, would have been credited.

This is the most difficult part of the question. Analyse information like this, and give good ideas associated with the problem. It was logical to suggest that the gene may reach another species and the consequences of this, e.g. the onset of sterility.

Exam practice questions

A *Answers on p. 30*

(1) **(a)** The drawing shows part of a DNA molecule.

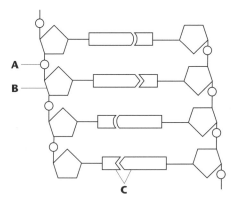

Name the parts labelled: **[3]**

(b) The text below shows the sequence of bases in a short length of mRNA.

A U G G C C U C G A U A A C G G C C A C C A U G

   (i) What is the maximum number of amino acids in the polypeptide for which this piece of mRNA could code? **[1]**

  (ii) How many different types of tRNA molecule would be used to produce a polypeptide from this piece of mRNA? **[1]**

 (iii) Give the DNA sequence which would be complementary to the first five bases in this piece of mRNA. **[1]**

(c) Name the process by which mRNA is formed in the nucleus. **[1]**

(d) Give **one** way in which the structure of a molecule of tRNA differs from the structure of a molecule of mRNA. **[1]**

[AQA B Specimen]

(2) The diagrams show six stages of mitosis, labelled A to F, in a plant root tip, as seen under high power of a light microscope.

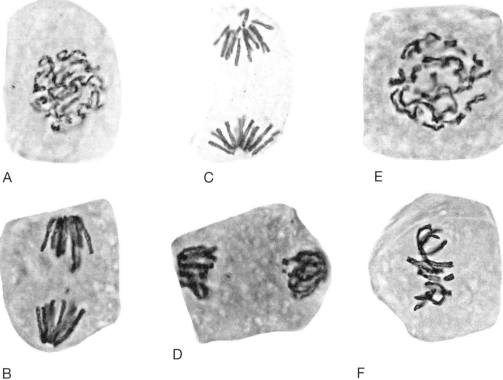

A   C   E

B   D   F

Exam practice questions

(a)   Name the stages of mitosis shown in the diagrams.   **[6]**

(b)   Explain the importance of mitosis to living organisms.   **[3]**

[OCR Specimen]

(3) Humans produce insulin from certain cells of the pancreas. The insulin gene is isolated from a human pancreas cell and then inserted into a plasmid. The DNA responsible for the synthesis of insulin is then inserted into a bacterium. The diagram, which is not drawn to scale, shows how insulin can be produced in this way. Different enzymes function at X and Y.

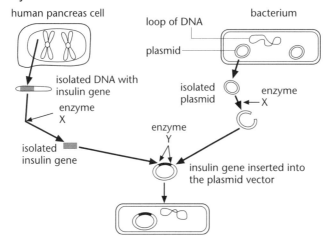

(a)   State a general term for the technique shown in the diagram.   **[1]**

(b)   Outline the roles of the enzymes that function at X and Y.   **[3]**

(c)   Explain why the plasmid is described as a vector.   **[2]**

(d)   Outline the role of the bacterium in the process once the vector has been
        inserted into the host cell.   **[4]**

[OCR Specimen]

Answers

(1)   **A** Phosphate   **B** Pentose/sugar/deoxyribose   **C** Bases/named bases

(b) (i) 8   **(ii)** 6   **(iii)** TACCG

(c)   Transcription

(d)   mRNA has codons, whereas each tRNA molecule has an anticodon.

Examiner's tip

If you cannot imagine the structure of each molecule try to draw out a quick diagram of each at the side of the paper. Detect the differences to score your marks.

(2) (a)   A Prophase   C Anaphase   E Prophase
               B Anaphase   D Late anaphase/early telophase   F Metaphase

Examiner's tip

LOOK FOR
Prophase: nucleus still visible, chromosomes have doubled.
Telophase: chromosomes have moved to the poles.
Anaphase: centromeres have just split, chromatids are beginning to move to the poles.
Late anaphase/early telophase: in transition between the two phases, chromatids moving to the poles.
Interphase: nucleus visible, chromosome are not visible.
Metaphase: chromosomes are at the middle of the cell, i.e. the equator.

(b)   The process produces genetically identical cells, for growth, for repair, for asexual reproduction.

(3) (a)   Genetic engineering/gene technology/gene manipulation/recombinant DNA technology.

Examiner's tip

The recommendation here is to give genetic engineering. 'Mainstream' answers are the safest responses. All Examination Boards have difficult decisions to make when producing their mark schemes. Some answers are only just acceptable. Always try to give a mainstream answer.

(b)   At X: cut the DNA open; cut at specific base sequences; cut to produce 'sticky' ends.
                                                              (any 1 answer)
               At Y: insulin gene attached to plasmid; to form a complete plasmid or ring of DNA; detail of recombination of DNA.   (any 2 answers)

(c)   The plasmid carries/transfers, gene/DNA, to another cell/to another bacterium/to another place.   (any 2 answers)

(d)   The bacterium multiplies the plasmid or clones the plasmid; the bacterium reproduces by cloning/bacteria multiply; insulin is secreted/insulin is produced; bacterium uses chemicals produced by its own metabolism; (insulin) produced by protein synthesis.   (any 4 answers)

Examiner's tip

Logically you should realise that once the gene has been transferred then both plasmid must be internally cloned and the bacterium itself, if large amounts of insulin are to be produced.

Questions with model answers

C grade candidate – mark scored 4/6

 For help see Revise AS Study Guide pages 96–99

(1) (a) The diagram below shows a germinating pollen grain and a mature ovule from an insect pollinated flower. Some nuclei have been labelled.

Examiner's Commentary

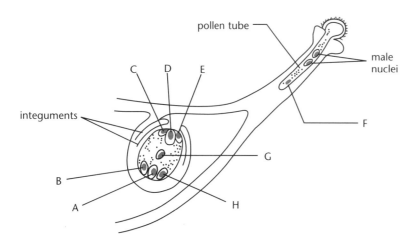

Give the letter of the nucleus which fuses with a male nucleus to form each of the following.   [2]

    The zygote   *A* ✗

    The endosperm   *G* ✔

The first letter is not correct. The candidate was aware that the middle nucleus is the 'egg nucleus' but chose the wrong end of the embryo sac. The egg nucleus is the middle of three at the micropyle end.

(b) Give **TWO** ways in which the structure of an insect pollinated flower differs from that of a wind pollinated flower of a grass. [2]

    1. An insect pollinated flower has colourful petals, whereas grass has green bracts instead ✔.

    2. An insect pollinated flower has a knob-like stigma whereas the grass has a feathery stigma structure ✔.

*Both answers correct. The candidate did well to mention **both** of the types of flower in each answer. If you stated that the insect pollinated flower has colourful petals there is no assumption that the other does not. Mention both then the answer is not ambiguous.*
Grass flowers have no sepals whereas insect pollinated flowers do.

(c) Describe **ONE** mechanism which prevents self-fertilisation in flowering plants.   [2]

    The pollen ripens and is shed before the ovules have matured ✔.

    This is called protogyny ✗.

[Edexcel Specimen]

*The candidate began well and described how, if the pollen ripens first, no self-fertilisation can be possible. The incorrect name of the process was chosen! **Protandry** is where **pollen matures first** whereas **protogyny** names the process whereby **ovules ripen first**. Better revision would prevent this type of error.*

A grade candidate – mark scored 15/15

For help see Revise AS Study Guide pages 103 and 104

(2) The table shows the concentration of some sex hormones in the blood of a cow over a period of time.

Time/days	Concentration of hormone in the blood/arbitary units		
	progesterone	oestrogen	LH
0	1	14	32
2	2	8	1
4	4	7	1
6	10	7	1
8	14	7	1
10	18	7	1
12	19	7	1
14	19	7	1
16	18	7	1
18	8	18	1
20	1	14	32
22	1	8	32
24	2	8	1

(a) Use the figures in the table to estimate the length of the cow's oestrous cycle. Explain how you arrived at your answer. **[2]**

20 days ✔

I worked it out by noting that the progesterone level of 1 unit took 20 days to return to the same level. There seemed to be a pattern ✔.

Correct. Each hormone shows a pattern such as LH was at 32 units at time 0 days and it took 20 days to return to the identical level of 32 units. Any hormone could have been given as an example.

(b) Explain how the high concentration of LH on day 0 caused an increase in progesterone in the days which followed. **[3]**

Luteinising hormone (LH) reached the ovary ✔, *this stimulated ovulation* ✔. *After ovulation the follicle changes and can secrete progesterone* ✔.

*All correct. LH travels to the ovary via the blood, stimulates ovulation then the follicle changes into a **corpus luteum**. This latter part could also have gained credit but the candidate already scored a maximum by referring to the progesterone which it secretes.*

(c) Progesterone is responsible for the growth of the lining of the uterus and the development of its blood supply. Suggest how and explain why the figures for progesterone would differ from those in the table if the cow had become pregnant. **[2]**

The level of progesterone would increase ✔, *because the lining of the uterus needs to be maintained through the pregnancy* ✔.

The corpus luteum does not break down so quickly when the cow is pregnant. Additionally the placenta secretes a lot more progesterone.

Questions with model answers

A grade candidate continued

 For help see Revise AS Study Guide pages 103 and 104

Examiner's Commentary

(d) The concentration of progesterone in milk can be measured. It gives a very early indication of whether or not a cow is pregnant.

(i) Suggest how progesterone gets into milk. **[1]**

It may diffuse from the blood ✔.

(ii) Explain why it is an advantage for a farmer to know as early as possible whether or not a cow is pregnant. **[2]**

A cow needs to have another calf to maintain maximum milk flow ✔.

If the farmer finds out she is not pregnant then he will have her mated at next ovulation ✔.

(e) Describe how hormones may be used as contraception and in controlling infertility in humans. **[5]**

Oestrogen is produced which inhibits the production of the hormone FSH ✔ ✔.

In this way the Graafian follicle does not develop, so oestrogen acts as a contraceptive ✔.

If the person is infertile due to a lack of FSH then supplying it will stimulate the development of follicles, so ovulation will then take place ✔ ✔.

[AQA A Specimen]

All correct. In part (i) it is logical to suggest diffusion from the blood. All hormones are transported via blood! In answer to part 2, an economic reason was needed. Milk volume decreases if no calf is produced, so the farmer aims for the production of a calf. Production of a calf, and the maintenance of milk are both economic reasons.

All correct! Progesterone could have been given but the candidate had already scored a maximum. Progesterone also inhibits FSH production and ensures that vaginal mucus is not the correct consistency for successful conception. The answers to this question part require continuous prose. Marking of the points would be based on the expression of logical points in clear scientific terms. The candidate would have also been credited with these communication marks.

Exam practice questions

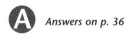 *Answers on p. 36*

(1) The diagram below shows a section through a human ovary.

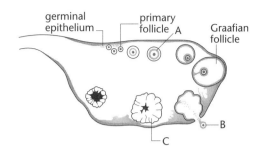

(a) Name parts A, B and C. **[3]**

(b) **(i)** Which part of the ovary divides to form the primary follicles? **[1]**

(ii) Which type of cell division is responsible for the production of the primary follicle? **[1]**

(c) FSH reaches the ovary so that part A begins to mature.

(i) Name the first hormone which is secreted by the ovary as a result of the arrival of FSH. **[1]**

(ii) Describe the role of this hormone in the menstrual cycle. **[1]**

(d) Structure B leaves the ovary.

(i) Where does structure B enter, immediately after leaving the ovary. **[1]**

(ii) Which hormone level peaks just before structure B leaves the ovary? **[1]**

(2) The diagram below shows four gametes produced from a cell **X**. Only one chromosome is shown in each nucleus. Different combinations of two pairs of alleles are shown along these chromosomes. **A** is dominant to **a**, and **B** is dominant to **b**.

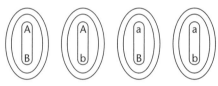

(a) Which type of cell division resulted in the production of these gametes from cell X? **[1]**

(b) Explain what, during this cell division, caused the production of the four *different* allele combinations? **[2]**

(c) Cell X, divided to form the gametes shown above. Cell X had previously been produced as a result of two processes, fertilisation, then cell division.

• The parental cell which produced the male gamete was homozygous dominant.

• The other parental cell which produced the female gamete was homozygous recessive.

• These gametes fused to form cell X.

(i) Which type of cell division followed fertilisation? **[1]**

(ii) Draw a labelled diagram of cell X to show the allele combination along the chromosomes in the nucleus. Use the information given in parts **(a)** and **(c)** to help you. **[3]**

Answers

(1) (a) A Primary oocyte

B Secondary oocyte

C Corpus luteum

(b) (i) Germinal epithelium

(ii) Mitosis

Examiner's tip

*The outer layer of cells (germinal epithelium) divides at the fetal stage to form oogonia. These go on to form primary follicles. Take care when giving answers about cell division. A key event in the ovary is meiosis but this only takes place as the **primary oocyte** divides. Before this stage divisions are mitotic!*

(c) (i) Oestrogen

(ii) It stimulates the build up of the endometrium/lining of the uterus.

Examiner's tip

Secreted into the blood the endometrium thickens and developes a capillary network in readiness for a potential embryo.

(d) (i) Fallopian tube/oviduct

(ii) Luteinising hormone/LH.

Examiner's tip

Fallopian tube should be easy to remember, but the hormone is more difficult. There are four important hormones controlling the menstrual cycle. Three of them were tested here.

(2) (a) Meiosis

(b) Cross-overs/chiasmata

and alleles are exchanged along the chromatids.

Examiner's tip

Remember that the chromatids cross at the equivalent locus or position. Here the alleles can be exchanged from one chromatid to another.

(c) (i) Mitosis

(ii) One chromosome has A + B

One chromosome has a + b

Two chromosomes in the nucleus

Questions with model answers

C grade candidate – mark scored 7/11

? *For help see Revise AS Study Guide pages 115 and 116*

(1) A diagrammatic representation of the nitrogen cycle is given below.

organic compounds
in plants

feeding

A

nitrogen
in air

F

via **E**

D

G

C

organic
compounds
in animals

defaecation

death

organic
compounds
in detritivores

decomposition

B

Examiner's Commentary

Bad start. The candidate gave the answers in the wrong positions!
*A – the only bacteria which can utilise atmospheric nitrogen are the **nitrogen fixing bacteria**, e.g. in the nodules of legumes like peas and beans.*
*D – are **nitrifying bacteria**. The clue here is that these bacteria act on the products of decomposition.*

(a) What is the descriptive term applied to organisms.   **[2]**

  (i)   A   *Nitrifying bacteria* ✗

  (ii) D   *Nitrogen fixing bacteria* ✗

(b) Name process B   **[1]**

    Excretion ✔

Correct. Death and defaecation were already given on the diagram. So the other process which contributes a product which can be decomposed is excretion.

(c) Name one group of organisms that can act as detritivores.   **[1]**

    Earthworms ✔

Bacteria and fungi would have been accepted.

(d) Name the chemicals C, F and E.   **[3]**

    C Ammonia ✔   *E Nitrites* ✔   *F Nitrates* ✔

(e) Explain G   **[2]**

    I think F is fertiliser so that G is an industrial process to make this fertiliser which uses atmospheric nitrogen ✔✗.

(f) The agricultural industry has been accused of causing eutrophication of lakes and rivers. By referring to the stages shown in the diagram explain how this may occur.   **[2]**

    Nitrate fertiliser (F) is produced ✗, *which runs off the fields into rivers and lakes* ✔.

*A further mark was available for an indication that the process produces artificial **nitrogenous fertiliser**.*

[WJEC Specimen]

Questions with model answers

A grade candidate – mark scored 9/12

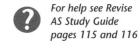

For help see Revise AS Study Guide pages 115 and 116

(2) The population dynamics of the great tit (*Parus major*) in a wood near Oxford has been intensely studied over the past 50 years. Population changes relate to the production and survival of young great tits, and these are influenced by:

- the availability of insect prey (e.g. caterpillars) for adults and young

- the predation (e.g. by sparrowhawks) of young birds.

Adult great tits pair in autumn and establish territories in the following January. The females lay a clutch of about ten eggs in late April. These hatch over a period of two or three days, nearly two weeks after the last egg has been laid. Both adults catch caterpillars for the nestlings, which take two weeks to grow feathers. They then begin to make practice flights. They are fed by the adults for a further two weeks. The period just after the young have left the nest is critical and serious predation by sparrowhawks may take place.

Figures 1 to 4 below relate to information concerning the number of eggs produced, the survival of the nestlings (young in the nest) and the survival of the fledgelings (young to three months after leaving the nest).

Summarise the findings of each figure, and using the information provided previously, suggest an explanation for each finding.

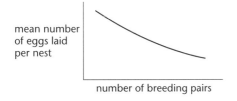

Figure 1   **[3]**

The average number of eggs per nest is greater, the fewer breeding pairs of birds there are in the wood ✔. The less breeding pairs there are in the wood the more food there is for the birds so they can produce more eggs ✔. When the breeding pairs are close together, then more energy is expended for aggressive behaviour so that fewer eggs are laid ✔. ← **All correct.**

Examiner's Commentary

A grade candidate continued

Examiner's Commentary

Figure 2 [3]

hatching success

number of eggs in nest

The less eggs in a nest the more successful is the hatching ✔.

The eggs in low numbers in nests may have more nutrients in the yolks so the chemicals are more suitable for development ✔.

Perhaps in nests with large numbers of eggs the incubation temperature cannot be maintained because outside eggs are not fully covered ✔.

All correct again! In ecological questions good ideas are needed. The first mark was straight forward, but reasons for the decrease in hatching success were needed. In questions like this try to picture the ecological situation as you make your analysis.

mean mass of nestlings

number of nestlings in nest

Figure 3 [3]

The more young there are in a nest the lower is the mass of the young birds ✔.

Parent birds collect food for the young, so the more there are to feed, the less each young bird is fed ✔.

.... ✗

A final mark was missed. One mark was available for the fact that there is less competition in a small family of young birds.

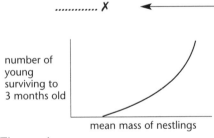

number of young surviving to 3 months old

mean mass of nestlings

Figure 4 [3]

The heavier the mass of the small birds, the more there are which survive ✔.

Maybe the larger young birds are better at avoiding the predators ✔.

.... ✗

A good start. The candidate unfortunately missed a mark. Bigger fledglings are less likely to starve. Additionallly the candidate should have mentioned the sparrowhawks as the predators in the second part.

[NICCEA Specimen]

Exam practice questions

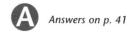 *Answers on p. 41*

(1) With respect to carbon dioxide, light and temperature as limiting factors in photosynthesis comment on the following agricultural practices,

    (a) Growing early crops under glass. **[2]**

    (b) Introducing additional CO_2 into commercial greenhouses when growing tomatoes. **[3]**

    (c) Painting domestic greenhouse glass with whitewash in the summer. **[2]**

[NICCEA Specimen]

(2) (a) **Circle** the ecological terms that can be used to describe human activities in the fishing industry.

        Density dependent factor Density independent factor Parasite Predator **[2]**

    (b) State **two** reasons why regulating the size of fishing net mesh size may **not** prevent over-fishing. **[2]**

    (c) Name the **two** biotic factors that tend to increase the numbers of an animal in a given area. **[2]**

    (d) Suggest **two** reasons why it would be difficult to gain an accurate estimate of these factors for a North Sea food fish. **[2]**

[WJEC Specimen]

Answers

(1) **(a)** Low temperatures limit photosynthesis in the cold months in the UK climate. Use of glass to cover plants, e.g. greenhouse warms up soil/or warms up atmosphere to increase the rate of photosynthesis.

Examiner's tip

An understanding of limiting factors is needed throughout this question. Glass always warms up the soil, above the ambient temperature. Photosynthetic rate is enhanced.

(b) An increase in CO_2 increases the rate of photosynthesis, over a range of different light intensities and different summertime temperatures.

Examiner's tip

Gas heaters are often used which increase temperature and give out CO_2.

(c) The greenhouses can heat up excessively due to sunshine, light is not a limiting factor in the summer.

Examiner's tip

Whitewash enables the greenhouse to remain cooler by reflecting some of the light. If it was all reflected then there could be no photosynthesis.

(2) **(a)** Density independent factor, Predator

Examiner's tip

Humans have a predatory role as fish are caught for food. Density independent factor is correct. Fishing aims to maintain the population beneath the carrying capacity.

(b) The number of fishing boats in an area may not be regulated/a lot of fishing boats may be still be allowed to fish; The time allowed for fishing may not be regulated; Each species may have an optimum net mesh size; Difficult to enforce.

(Any 2 accepted)

Examiner's tip

Always try to think logically around the problem. Net size is only one factor. The frequency of fishing and numbers of fishing vessels are both key factors.

(c) Birth rate, Immigration

Examiner's tip

Biotic factors are associated with the organisms in an area. The Examination Board may have credited two other responses. The numbers of predators in an area or the numbers of prey may influence the population of fish.

(d) Counting the number of females laying eggs is very difficult; Counting the number of eggs which actually hatch is also a very difficult task; Surveying the distribution of fish populations is very difficult as they are motile; Surveying the populations to find out the age structure of the population is very difficult; The area is a huge size to survey.

(Any 2 points)

Examiner's tip

All of the above factors could be applied to any biotic factor given in answer to part (c). Remember that in ecological type questions you will need lots of good logical ideas. At AS level you are expected to demonstrate analytical skills and apply them to new situations.

Questions with model answers

C grade candidate – mark scored 6/10

? For help see Revise
AS Study Guide
pages 130–136

Examiner's Commentary

(1) **(a)** Explain the meaning of the following terms as they apply to infectious diseases: **[2]**

Endemic – *this means that the disease is always present in the population, such as malaria in the population of many African nations* ✔

Epidemic – *this is where a disease has spread throughout the population of an area, e.g. a city* ✔.

The table below shows the diseases which cause death in developing and developed countries.

developing countries		developed countries	
disease	percentage deaths	disease	percentage deaths
diarrhoea	42	heart diseases	32
respiratory infections: e.g. tuberculosis (TB)	25	cancers	23
malnutrition	10	strokes	12
malaria	7	bronchitis	6
measles	15	pneumonia	5
others	11	others	22

(b) With reference to the table above.

(i) Explain why infectious diseases are leading causes of death in developing countries. **[4]**

The food situation is not good ✗*. They have poor living conditions* ✗. *In the developing countries there may be no system of good sanitation so that diseases are spread via water* ✔.

They cannot afford vaccinations ✔.

The first two responses did not justify a mark. The people often suffer from **malnutrition**. *This reduces the effectiveness of the immune system. These are worthy of credit. Reference to the food situation not being good is much too vague.*
Similarly having poor living conditions is vague. What does this tell us? 'Poor living conditions encourage the spread of diseases' would have gained credit. Remember to give detail!

(ii) Explain why degenerative diseases are leading causes of death in developed countries. **[4]**

Many people in the developed countries are better off ✗.

A lot of people are overweight ✗.

The countries can afford to give the people vaccinations ✔.

Diseases like cancer take a longer time to develop and are not noticed until the disease is well established ✔.

The fact that developed countries have economic advantages is too vague. This could be linked to the fact that infectious diseases are controlled. 'Overweight' is another vague term. Obesity leading to a circulation problem, such as atherosclerosis or heart attack, gives the detail expected at this level.

[OCR Specimen]

A grade candidate – mark scored 17/19

For help see Revise AS Study Guide pages 130–136

Examiner's Commentary

(2) The bacterium, *Vibrio cholerae*, is the causative agent of cholera. The El Tor strain of *V. cholerae* originally occurred only in Indonesia. In 1961, this strain began to spread replacing existing strains in other parts of Asia. El Tor is now widespread throughout Asia, the Middle East, Africa and parts of Eastern Europe, but has never established itself in Western Europe.

El Tor is hardier than the strain it replaced and the bacteria may continue to appear in the faeces for up to three months after patients have recovered. The bacteria may persist in water for up to fourteen days.

(a) State **two** ways in which *V. cholerae* is transmitted from infected to unaffected people. **[2]**

> 1. *It is carried by water, such as irrigation water on vegetables* ✔.
>
> 2. *Via faeces of an infected person* ✔. ◄

Both correct. Additional credit could have been given for transmission via food, drinking water or not washing hands after using the toilet or even carried by flies to food.

(b) Suggest how laboratory tests could identify carriers of cholera. **[2]**

> *Some people infected with cholera have mild symptoms or none at all, and are carriers of the disease.*
>
> *Grow the bacteria from a sample of the person's faeces* ✔. ◄
>
> *Identify the bacteria* ✔.

Correct but there were other possible answers. Test for antibodies against V. cholerae, use of a microscope, use of monoclonal antibodies.

(c) Suggest **four** reasons why El Tor has not become established in Western Europe. **[4]**

> 1. *Sewage treatment takes place so that sanitation is good* ✔.
>
> 2. *So water is not contaminated* ✔.
>
> 3. *Water is purified with the help of chlorine to kill bacteria* ✔.
>
> 4. *................* ✗ ◄

The candidate ran out of ideas after giving the first three correct responses. This example illustrates a good examination tip. A candidate could be tempted to give one answer about water, but here two are credited. Give full detail to gain full credit.
*The Examination Board **rejected** vaccination or use of antibiotics.*

The United Nations, recognising that most of the outbreaks of cholera were the result of polluted water supplies, set up a 'Decade of Water' in 1981. Its aim was to provide safe water for everyone. Over the decade 1981/1990, the number of people lacking a safe water supply in developing countries dropped from 1800 million to 1200 million.

Questions with model answers

A grade candidate continued

For help see Revise
AS Study Guide
pages 130–136

Examiner's Commentary

(d) Explain why cholera continues to be a worldwide problem, in spite of the 'Decade of Water' campaign. **[8]**

(In this question, 1 mark is available for the quality of written communication.)

The cholera bacterium is passed via polluted water. Many people still do not have a supply of clean water ✔. Only 600 million people out of the 1800 million people were given access to a safe water supply, leaving 1200 million people vulnerable ✔. These people did not have access to clean water ✔. The programme is not 100% effective. In fact it is around 66.6% effective ✔.

One of the problems may be that areas are large and rural. They are difficult to reach and supply with the safe water supplies ✔. Additionally, the population in the area is increasing, so there is more chance of the diseases being passed on ✔.

Suitable housing is not always available so that shanty towns develop with poor sanitation ✔.   (Communication mark) ✔

Full marks would be awarded. The last sentence has two valid points, i.e. the build up of shanty towns, and poor sanitation. The candidate reached the maximum for this part of the question and could not gain further credit.
*A further mark would have been credited for **written communication**. The answer satisfied these criteria, 'legible text with accurate spelling, punctuation and grammar'.*

The antibiotic tetracycline is sometimes used as a treatment for cholera.

(e) (i) Suggest **two** ways in which tetracycline can affect *V. cholerae*. **[2]**

It is bacteriostatic, preventing further growth of bacterial cells ✔.

Specifically it prevents protein synthesis ✔.

Full marks but others were available, such as the tetracycline causing the lysis or splitting open of the cell membrane.

(ii) Explain why tetracycline should not be used routinely for all cases of cholera. **[1]**

Bacteria become immune to the antibiotic ✗.

*The candidate is close, but would not be awarded credit. The bacteria can become **resistant** to the antibiotic. Immune is not an equivalent term.*

[OCR Specimen]

Exam practice questions

A *Answers on p. 46*

(1) (a) What is a disease? **[1]**

(b) Describe FOUR different types of disease. Give a specific example of each type of disease in your answer. **[8]**

(c) What is the meaning of the term pandemic? **[1]**

(2) (a) Read the following passage on the disease malaria, and then write in an appropriate word or words on each dotted line to complete the passage.

The disease, malaria, is caused by the protozoan parasite, ... This dangerous parasite is carried by a .... This is a .... mosquito. The mosquito may feed on a person who may is suffering from malaria. It does this at night by inserting its .... into a .... beneath the skin. The digestion of the mosquito releases the malarial parasites which burrow into the insect's stomach wall and breed there. Some move to the .... Next time the mosquito feeds it secretes saliva to prevent clotting of the blood. This introduces the .... into the person's blood, who is likely to contract the disease. **[8]**

(b) Describe FIVE different ways of preventing malaria. **[5]**

Answers

(1) (a) A disease is a disorder of a tissue, organ or system of an organism. As a result of the disorder symptoms are evident.

(b) Infectious disease by pathogens which attack an organism and can be passed from one person to another, e.g. infectious disease to include measles, etc.
Genetic diseases (congenital diseases) can be passed from parent to offspring, e.g. any genetic disease to include haemophilia, cystic fibrosis, etc.
Dietary related diseases, as a result of the foods which we eat. Too much or too little food may cause disorders, e.g. obesity/anorexia nervosa/deficiency diseases, e.g. lack of vitamin D causes the bone disease rickets.
Environmentally caused diseases where some aspect of the environment disrupts bodily processes, e.g. As a result of nuclear radiation leakage, cancer may result.
Auto-immune disease where the body, in some way attacks its own cells, so that processes fail to function effectively, e.g. leukaemia where phagocytes destroy a person's own red blood cells.

(Any 4 answers)

(c) This refers to an outbreak of a disease over a very large area, such as a continent.

(2) (a) The disease, malaria, is caused by the protozoan parasite, **Plasmodium**. This dangerous parasite is carried by a **vector**. This is a **female Anopheles** mosquito. The mosquito may feed on a person who may is suffering from malaria. It does this at night by inserting its **stylet** into a **blood vessel** beneath the skin. The digestion of the mosquito releases the malarial parasites which burrow into the insect's stomach wall and breed there. Some move to the **salivary glands**. Next time the mosquito feeds it secretes saliva to prevent clotting of the blood. This introduces the **parasites** into the person's blood, who is likely to contract the disease.

Examiner's tip

*This is known as a cloze exercise. Knowledge is required but the clues are there to help you. Logically the first word is Plasmodium, the correct term for the malarial parasite. The tricky marks are the two given for **female**, and **Anopheles** mosquito. The key skill here is to remember the detail.*

(b) Insecticide sprayed onto lake surfaces kills mosquito larvae/oil poured on lake surfaces prevents air entering the breathing tubes of the (mosquito) larvae so they die.
Fish can be used as predators in lakes to eat the larvae/use biological control.
Drain ponds to remove the mosquitoes' breeding area/cover up all waste containers to remove the breeding area.
Use *Bacillis thuringiensis* to destroy mosquitoes.
Mosquito nets exclude mosquitoes from buildings/from beds.
Electronic insect killer attracts the mosquitoes via ultra-violet light then kill them by application of voltage.
Use of drugs to kill the malaria organism in sufferers/isolate the people suffering from malaria.

(Any 5 answers)

Questions with model answers

C grade candidate – mark scored 3/5

*For help see Revise
A2 Study Guide
page 28*

Examiner's Commentary

(1) The diagram shows the main stages in the light-independent reactions in photosynthesis.

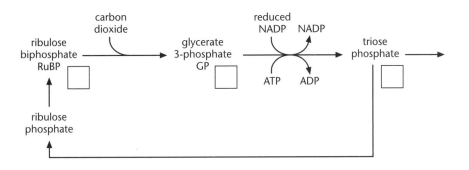

(a) Write in the boxes in the diagram the number of carbon atoms in each of the relevant substances. **[1]**

    5   6 ✗   3 ◄───

Incorrect. The middle value is the wrong number of carbon atoms. The candidate realisd that there were 6 carbon atoms, 5 from RuBP and 1 from carbon dioxide, but 2 × 3-carbon glycerate 3-phosphate molecules are formed.

(b) What is the role of ATP in the conversion of:

(i) Glycerate 3-phosphate to triose phosphate. **[1]**

It transfers energy for the reaction to take place ✔. ◄───

The break down of ATP always releases energy for many processes vital to life.

(ii) Ribulose phosphate to ribulose biphosphate. **[1]**

It supplies another phosphate group to the ribulose phosphate ✔. ◄───

Additionally the ATP supplies energy in the conversion.

(c) A plant was allowed to photosynthesise normally. The light was then switched off. Explain why there was a rise in the amount of glycerate 3-phosphate present in the chloroplasts of this plant. **[2]**

NADPH$_2$ is produced in the light dependent reaction, so it is no longer being produced ✔.

It is needed to convert glycerate 3-phosphate ✗. ◄───

*The second part of the answer does not give the essential detail that glycerate 3-phosphate is no longer converted to **triose phosphate**. Without this credit was not available.*

[AQA A Specimen]

Questions with model answers

A grade candidate – mark scored 10/12

For help see Revise A2 Study Guide page 28

(2) The diagram shows some of the reactions which occur during aerobic respiration in an animal cell.

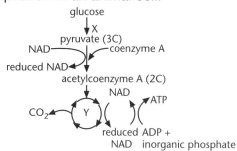

Examiner's Commentary

(a) (i) Identify the pathways X and Y. **[2]**

X Glycolysis ✔ Y Kreb's cycle ✔

Glucose to pyruvate is the key information to indicate glycolysis. After acetyl co-enzyme A the cycling of substances is clearly Kreb's cycle. TCA or citric acid cycle are equivalent to Kreb's cycle.

(ii) State precisely where pathway X occurs. **[1]**

Cytoplasm ✔

Cytosol would have been credited.

(b) Explain why one of the enzymes involved in the conversion of pyruvate to acetyl coenzyme A is called pyruvate dehydrogenase. **[2]**

The enzyme removes hydrogen ✔ from pyruvate ✔ and the hydrogen is accepted by acetyl co-enzyme A.

The final part of the answer was also correct but the candidate had already achieved the maximum.

(c) State what is meant by the term decarboxylation. **[1]**

This is the removal of carbon dioxide ✔.

Correct! Decarboxylation is the removal of carbon dioxide in a reaction.

(d) (i) State the site of oxidative phosphorylation in an animal cell. **[1]**

Mitochondria ✗

*Not enough! The answer required was the **inner membranes** of the mitochondria.*

(ii) Describe in outline the production of ATP in oxidative phosphorylation. **[5]**

Hydrogen molecules are passed along the inner membrane of the mitochondria ✔. Electrons are passed along a series of carriers ✔. As electrons passed from one carrier to another, the molecule losing an electron is oxidised and the molecule accepting the electron is reduced ✔. At the end of the process oxygen is the final electron acceptor ✔. .... ✗

[OCR Specimen]

The candidate fell short of full credit. ATP synthetase is involved in ATP production. This would have gained credit. Additionally the final mark could have been gained by referring to the proton pump in the inner membrane.

Exam practice questions

A *Answers on p. 50*

(1) The diagram below summarises the biochemical pathways involved in photosynthesis.

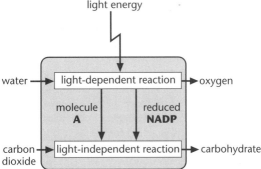

(a) Name Molecule A. **[1]**

(b) **(i)** Describe how NADP is reduced in the light-independent reaction. **[2]**

(ii) Describe the part played by reduced NADP in the light-independent reaction. **[2]**

[AQA B Specimen]

(2) **(a)** The diagram below shows the apparatus used to measure the effect of the light intensity on the rate of photosynthesis in *Elodea*.

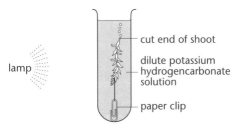

(i) Explain why it is possible to study the rate of photosynthesis using the above apparatus. **[1]**

(ii) Explain why potassium hydrogencarbonate was used. **[1]**

(iii) Give **one** reason why the results using this apparatus may not be accurate. Explain your answer. **[2]**

(b) The graph shows the uptake and release of carbon dioxide which takes place in a plant during a twelve hour period.

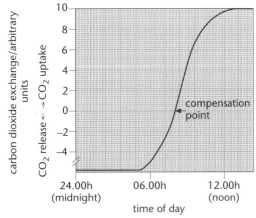

(i) At the compensation point carbon dioxide is neither taken up nor released. Suggest an explanation for this observation. **[1]**

(ii) This plant increased its mass only after the compensation point had been reached. Use the information in the graph and your own knowledge to explain the reason for this. **[3]**

[WJEC Specimen]

Answers

(1) (a) ATP

(b) (i) Electrons are raised to a higher energy level.
They are passed through a series of hydrogen acceptors.
Protons/H^+ ions are produced from the photolysis of water,
these go on to reduce the NADP. (Any 2 answers)
(ii) Reduces,
glycerate 3-phosphate/GP

Examiner's tip

The diagram given should be similar to the one you learned for your course. It is necessary to make the link between the way in which a process is presented in the examination and the way that you have learned it during your course.

(2) (a) (i) The rate of oxygen production is proportional to the rate of photosynthesis.

Examiner's tip

Note the level of response demanded at A2 level. You may have been tempted to write that oxygen is given off. It is the rate per unit time that is required and the idea that O_2 production is proportional.

(ii) To make sure that carbon dioxide is not a limiting factor.
(iii) Reason: heat from the lamp could increase the water temperature.
Explanation: this could affect the enzymes involved in photosynthesis.
OR
Reason: cannot count bubbles accurately.
Explanation: bubbles may be released very quickly.

Examiner's tip

These answers are linked. This means that if a reason is given, the wrong explanation would not be credited. Additionally the Examination Board informed examiners to expect other sensible reason-explanation pairs. Always try to think of logical ideas.

(b) (i) Rate of photosynthesis equals the rate of respiration.

Examiner's tip

It could have been stated as 'the amount of carbon dioxide produced in respiration is exactly taken up by photosynthesis'.

(ii) Respiration involves the break down or use of organic substances or starch.
When photosynthesis is greater than respiration there is an increase in glucose/starch/organic substances.
Organic substances accumulate,
which can be used to produce new cells. (Any 3 accepted)

Examiner's tip

After the compensation point is exceeded then photosynthesis will increase the mass of a plant. The greater the rate after this point the greater the growth rate.

Questions with model answers

C grade candidate – mark scored 7/10

For help see Revise AS Study Guide page 37

(1) The diagram of an electron micrograph shows some of the cells which form the lining of the mammalian small intestine.

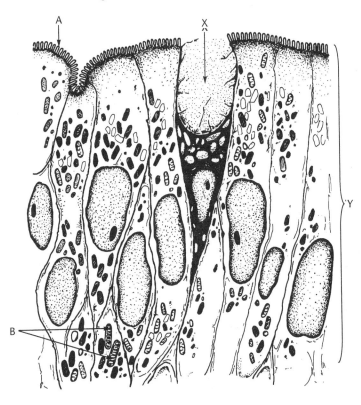

Examiner's Commentary

(a) What is the **general** name given to a tissue such as that labelled **Y**? **[1]**

　　　Columnar epithelium ✔

*Correct. The candidate gave too much information! The general term needed was **epithelium**.*

(b) (i) Name the features labelled A and B in the diagram. **[2]**

　　　A Villi ✗　*B Mitochondria* ✔

A is not correct. It is a big mistake! The answer should be microvilli or brush border. Villi consist of many cells whereas it is microvilli that are situated along the cell surface membrane.

(ii) Explain **fully** the function of **A** and **B** in this tissue **[4]**

　　　A Gives a very large surface area, and ✔ *gives an increased amount of absorption* ✔.

　　　B This releases energy ✔ .... ✗

The candidate began well but failed to give a function for the energy release. It is needed for the active transport of some food molecules through the cells into the blood.

(c) (i) Name the secretion labelled **X**. **[1]**

　　　Mucus ✔

(ii) State **two** functions of this secretion. **[2]**

　　　It is a lubricant ✔ *so that food can move through the intestines easily* ✗.

*The second part of the answer merely describes the function of a lubricant. It repeats the first idea so there can be no further credit. '**Protection against digestion by protease enzymes**' and '**protects against acid from stomach**' could have gained the final mark.*

[WJEC Specimen]

Questions with model answers

A grade candidate – mark scored 12/13

? *For help see Revise A2 Study Guide page 37*

Examiner's Commentary

(2) Different concentrations of maltose were placed in the small intestine of a mammal. The amounts of glucose appearing in the blood and in the small intestine were measured. The results are shown in the graph.

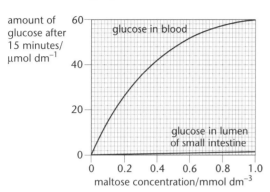

(a) (i) Give the name of the blood vessel most likely to have been sampled for glucose. **[1]**

Hepatic portal vein ✔

(ii) By what chemical process is a molecule of maltose converted into two molecules of glucose? **[1]**

Digestion ✗

No mark awarded! This question is at A2 Level and the required answer is Hydrolysis. All digestive enzymes are able to hydrolyse.

(b) The enzyme maltase converts maltose into glucose. This enzyme is found in the cell surface membrane of the epithelial cells of the small intestine.

(i) Explain the evidence from the graph which supports the view that the breakdown of maltose does not occur in the lumen of the small intestine. **[2]**

Very little increase in glucose in the lumen of the intestine ✔*, if the break down of the maltose took place here, glucose would increase in this position* ✔*.*

Additionally, the fact that it would take time for the glucose to diffuse through the intestine wall, could have been credited but the candidate had already achieved a maximum of two marks.

(ii) Suggest an explanation for the shape of the curve showing the change in the amount of glucose in the blood. **[3]**

As maltose concentrations increase more (maltase) active sites become occupied producing an increasing amount of glucose in the blood ✔*.*
The graph begins to level off because all of the active sites are occupied, so the enzyme limits the reaction ✔✔*.*

The candidate showed understanding of the enzymic action taking place and the consequence of all active sites being involved at a specific maltose concentration. Do not be tempted to discuss rate of reaction here. No time is mentioned!

A grade candidate continued

For help see Revise
A2 Study Guide
page 42

Examiner's Commentary

A cow obtains most of its nutritional requirements from mutualistic microorganisms in its rumen. The diagram summarises the biochemical processes carried out by these microorganisms.

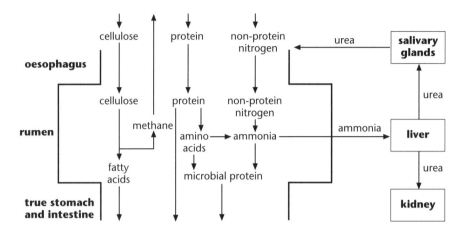

(c) Use the information in the diagram to help explain why:

(i) the relation between the cow and the microorganisms which live in its rumen may be described as mutualistic. **[2]**

The cows obtain fatty acids because the microorganisms are present ✔. The microorganisms gain cellulose and urea from the cow ✔.

Mutualism is the term that is used to describe organisms that live together where each gives an advantage to the other. In a mutualism question give one advantage for each of the partners in the relationship.

(ii) it is possible for a cow to survive on a diet which is poor in protein. **[2]**

Microorganisms use non-protein nitrogen ✔, which they convert into their own type of protein (microbial protein) ✔.

During your course you may not have encountered all of the details given in the flow diagram. Do not worry about this! Analyse the diagram, use the facts you do know, and logical answers will appear in the given information.

(iii) ruminant animals such as cows are less efficient than non-ruminant animals in converting energy in food into energy in their tissues. **[2]**

The microorganisms utilise some of the energy from the food ✔, some energy is lost by the production of methane ✔.

[AQA A Specimen]

Always remember the key information about ruminants: their microbial allies which live in the rumen. The microorganisms have an energy requirement!

Exam practice questions

 Answers on p. 55

(1) (a) Hormones are involved in the control of digestion. Complete the following table by writing the missing information in the empty boxes.

Stimulus	Endocrine gland	Hormone	Effect
food in stomach	stomach mucosa		stimulates oxyntic cells to secrete HCl
food in ileum		villikinin	stimulates the muscle in villi to contract rhythmically,
HCl in ileum	intestinal mucosa	secretin	1 2
Peptides and dipeptides in ileum	intestinal mucosa		stimulates the pancreas to produce enzymes

[5]

(b) The gall bladder contracts to expel bile into the bile duct.
  (i) Name the hormone which causes the gall bladder to contract. [1]
  (ii) Explain the roles of bile in the small intestine. [4]

(2) The figure below shows a longitudinal section of a villus, from the small intestine, highly magnified.

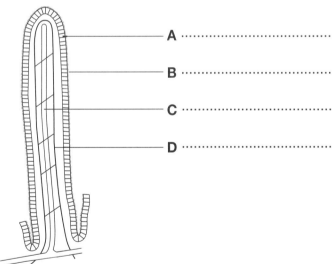

A ....

B ....

C ....

D ....

(a) (i) On the figure label structures **A** to **D**. [4]
  (ii) Put a ring around the figure below which is the most appropriate for the actual length of a villus. [1]
       2.5 mm   15 mm   2.5 μm   0.75 mm   250 μm
(b) (i) State in which tissue you would expect to find goblet cells [1]
  (ii) Explain their function [1]
(c) Explain how the products of digestion pass across the outer membrane of the cells labelled A in the figure.
(In this question one mark is available for the quality of written communication.) [9]

[OCR Specimen]

Answers

(1)
(a)

Stimulus	Endocrine gland	Hormone	Effect
food in stomach	stomach mucosa	**gastrin**	stimulates oxyntic cells to secrete HCl
food in ileum	**intestinal mucosa**	villikinin	stimulates the musclein villi to contract rhythmically,
HCl in ileum	intestinal mucosa	secretin	1 **stimulates pancreas to secrete more water** 2 **stimulates pancreas to secrete more alkali (ions)**
Peptides and dipeptides in ileum	intestinal mucosa	**pancreozymin**	stimulates the pancreas to produce enzymes

Examiner's tip

The task was to complete the gaps in the table. You need to recall information and are helped by the cues in the table. Moving across and down the table will help you to pin point the missing information. Checking out a table in this way unlocks your memory. Without revision the technique does not work!

(b) (i) Cholecystokinin
(ii) Bile is alkaline, helps neutralise acid from the stomach, bile emulsifies fats/bile emulsifies oils, makes a high surface area of fat or oil globules or chylomicrons, so that lipase can work effectively.

Examiner's tip

Never state that bile breaks down fats. It does not! After bile action the fats are merely broken up into tiny fat globules. High surface area of fats aids the action of lipase.

(2) (a) (i) A Columnar epithelium cells, B Microvilli/brush border,
C Lacteal/lymphatic vessel, D Blood capillary.
(ii) 0.75 mm
(b) (i) Epithelial layer of cells **(ii)** Secrete mucus which is alkaline.
(c) Quality of written communication is assessed in this answer.
One mark for legible text with accurate spelling, punctuation and grammar.

diffusion;	example;
facilitated diffusion;	
reference to the use of carrier molecules/carrier proteins;	example;
co-transport with Na^+;	example;
active transport;	
reference to ATP;	
reference to gradients;	example (such as glucose transport across membrane);
reference to membrane pumps;	
reference to selective uptake;	example;
pinocytosis;	(Any 8)

Examiner's tip

*There are 15 possible answers but you can score up to a maximum of 8. If you could remember the three mechanisms of **how** substances cross membranes then you could answer. Referring to the gradient, e.g. low to high concentration for active transport gives additional credit.*

Questions with model answers

C grade candidate – mark scored 3/5

 For help see Revise A2 Study Guide page 56

(1) (a) Describe the role played by calcium in muscle contractions. **[2]**

Calcium ions, are released from the sarcoplasmic reticulum **✗**.

They remove the blocking molecules from the binding sites on the actin filaments so that contraction can take place **✔**.

(b) The diagrams show the appearance of the sarcomere at points **A** and **B**.

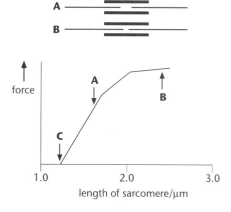

(i) What is the relationship between the force generated by the muscle and the number of actomyosin bridges? Explain your answer. **[2]**

Relationship - the more cross bridges there are the greater the force generated **✔**.

Explanation - the cross bridges become shorter and the actin and myosin filaments pull over each other, which generates the force **✔**.

(ii) Draw a simple diagram in the space below to show the appearance of the sarcomere at point C on the graph. **[1]**

the actin filaments meet **✗**

[AQA B Specimen]

Examiner's Commentary

The first part is true but it does not answer the question. The mark scheme gave 'the initiation of contraction'. The credited answer is seen as equivalent to this. The candidate failed to state the formation of the actomyosin bridge.

Correct. The candidate demonstrated understanding of the sliding filament hypothesis and the 'ratchet mechanism'.

The candidate considered that the actin filaments meet in the middle. Not correct! The actomyosin cross bridges pull the actin filaments across each other.

A grade candidate – mark scored 14/15

For help see Revise A2 Study Guide page 68

(2) The diagram shows some important features of homeostasis in the body.

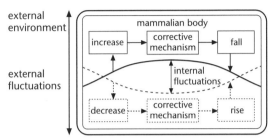

external environment

external fluctuations

mammalian body

increase → corrective mechanism → fall

internal fluctuations

decrease ---→ corrective mechanism ---→ rise

(a) Use the information in the diagram to help explain the importance of a mammal maintaining a constant internal temperature. **[4]**

The diagram shows that the body temperature does vary, but within a narrow range at either side of normal body temperature ✗.
A temperature too high above normal body temperature would denature the enzymes ✔. A temperature too far below normal body temperature would slow down the enzymic reactions ✔.
Substances produced during reactions will not be in balance ✔.

3/4 marks. The first part was almost correct. The candidate ignored information about the environmental temperature fluctuations. Credit was available for the fact that body temperature fluctuates less than the temperature of the external environment. Always remember that maintaining body temperature allows a mammal to live in different places. This would have gained credit.

(b) Explain the role of the hypothalamus and nervous system in the regulation of body temperature. **[5]**

Body temperature is monitored by receptors in the blood vessel leading to the hypothalamus ✔✔. Thermoreceptors in the skin result in action potentials being sent to the CNS (central nervous system) ✔. If the blood temperature is too high then the skin arterioles constrict. This is vasoconstriction ✔. So less heat radiates from the skin, as blood is diverted to the body core ✔

All correct! A maximum was achieved but a mark was available for the following principle that control involves receptor, hypothalamus and an effector.

(c) Explain why, in a normal healthy individual, the blood glucose level fluctuates very little. **[6]**

The level of blood sugar (glucose) is regulated by the hormones ✔. When the glucose in the blood increases, then more insulin is produced by the pancreas ✔. The increase allows more glucose into cells ✔. Additionally the insulin allows glucose into the liver to be converted into glycogen ✔. When the glucose in the blood decreases then the pancreas secretes glucagon ✔✔.

[AQA A Specimen]

*Full marks. When there are six marks available then you should make six clear points. It appeared that the final statement scored two marks. This is not so! It merely confirmed that the candidate was aware that an increase of glucose results in insulin secretion **and** lower glucose results in glucagon secretion.*

Exam practice questions

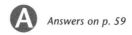 *Answers on p. 59*

(1) The diagram below shows a rod cell from the retina of a mammal.

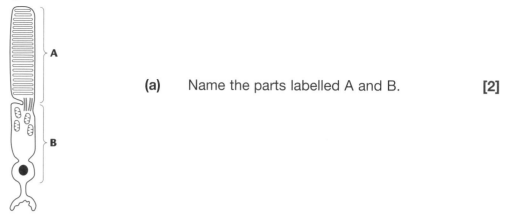

    (a)   Name the parts labelled A and B.   **[2]**

    (b)   State the location of most of the rod cells in a human retina.   **[1]**

    (c)   Give the name of the light sensitive pigment contained in the rod cells.   **[1]**

    (d)   Use the letter P to label on the diagram the region of the rod cell in which this pigment is located.   **[1]**

[Edexcel Specimen]

(2) Read the following passage.

Arrow poisons were widely used for hunting in South America where most of the powerful poisons, such as curare, originated. Charles Waterton carried out experiments on curare in the 1920s. He gave curare to a donkey which appeared to die ten minutes later. The animal was then revived by artificial ventilation of its lungs
5 with a pair of bellows and went on to make a full recovery. The experiment showed that injection of arrow poison into the blood stream causes death by respiratory failure. It is now known that curare competes with acetylcholine molecules for receptrons at neuromuscular junctions.

Since Waterton's work, many other chemicals have been discovered which also
10 affect the nervous system, two of which are anatoxin and saxotoxin.
Anatoxin also affects synapses. Its molecules are very similar to those of acetylcholine but they are not broken down by the enzyme, acetylcholinesterase. Saxotoxin is quite different and blocks the sodium channel proteins in nerve axons.

    (a) During synaptic transmission, acetylcholine is released from the presynaptic neurone into the synaptic cleft.

      (i) Describe how an action potential brings about the release of acetylcholine into the synaptic cleft.   **[2]**

      (ii) Describe how this acetylcholine may cause a new action potential to develop in the postsynaptic nerve cell.   **[3]**

    (b) Explain how injection of arrow poison into the blood stream causes death by respiratory failure (lines 6–7).   **[3]**

    (c) **(i)** Explain how acetylcholinesterase is important in the functioning of a synapse.   **[2]**

      (ii) Suggest an explanation for the fact that one of the symptoms of anatoxin poisoning is the excessive production of tears.   **[3]**

    (d) Explain how saxotoxin may cause paralysis.   **[2]**

[AQA B Specimen]

Answers

(1) (a)   A Outer segment   B Inner segment

Always remember the direction that light hits the retina, from label B to A.

(b)   Outside of the fovea/not in fovea

The fovea contains mostly cones and, correspondingly, most rods are outside this position.

(c)   Rhodopsin
(d)   Label P on any one membrane in the outer segment.

The horizontal membranes in the outer segment contain the pigment.

(2) (a) (i) Calcium ions enter the pre-synaptic membrane, the vesicles fuse with the membrane, releasing their contents/releasing acetylcholine.
  (ii) (Acetylcholine molecules) diffuse across cleft, they fit into receptor sites/receptor molecules/receptor proteins in the post synaptic membrane. This opens channel proteins, Na$^+$ ions flow into the post synaptic neurones.

If a threshold is reached then an action potential in the post synaptic cell results. Not on the mark scheme of the Examination Board it is likely that the final statement would be credited. If a threshold value of Na$^+$ ions is not reached then an action potential is not elicited.

(b)   The drug binds with the receptor molecules at the neuromuscular junctions, acetylcholine cannot fit, so that the muscles cannot contract, so breathing is not possible.   (Any 3 marks)
(c) (i) Acetylcholinesterase breaks down acetylcholine, so it prevents further action potentials/prevents further impulses.

The acetylcholinesterase ensures the removal of acetylcholine from the receptor sites so that the sodium channels close. Further action potentials are prevented in this way.

  (ii) The production of tears is controlled by the parasympathetic nerves, which have cholinergic synapses.
Anatoxin remains in the acetylcholine receptor sites and is not able to be broken down, so impulses continue to be sent.   (Any 3)
(d)   No sodium ions can enter the neurone, so no action potential/impulse can be produced.

From information in the passage it seems that the substance blocks the channel proteins directly. With no action potentials in the nerve cells, no muscle contracts, so paralysis is the result.

Questions with model answers

C grade candidate – mark scored 4/5

 For help see Revise A2 Study Guide pages 79 and 85

(1) A queen honey bee can lay both fertilised and unfertilised eggs. Fertilised eggs develop into diploid females and unfertilised eggs develop into haploid males. The diagram shows the formation of gametes in female bees and in male bees.

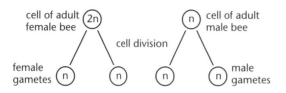

Examiner's Commentary

(a) Giving a reason for your answer in each case, name the type of cell division in the bee that produces: [2]

  (i) female gametes

   This type of cell division is meiosis, the reason for this is that the number of chromosomes is halved ✔. ◄

  (ii) male gametes.

   This type of cell division is mitosis, the reason is that the number of chromosomes stays the same as in the parent cell ✔.

Both correct. The key information given was that the parent cell was 2n, and the resulting cells from division were haploid (n). This is meiosis! Correspondingly a cell which is 'n' produces more 'n' cells. This is mitosis!

(b) The table shows some features which contribute to variation in the offspring of bees. Complete the table with a tick if the feature may contribute or a cross if it does not.

Feature	Female offspring	Male offspring
Crossing over	✔	✗
Independent segregation of chromosomes	✔	✗
Random use of gametes	✔	✗

                                   [2]

*The first column is correct. The candidate identified the female as being produced by meiotic division. During meiosis crossing over takes place **and** independent segregation of chromosomes. Then, since different gametes fuse, the final box can be ticked. In the second column there should be two ticks then a cross. The queen produces the unfertilised eggs which give rise to the males. These eggs must have undergone crossing over and independent segregation.*

(c) Body colour in bees is determined by a single gene. The allele **B** for yellow body is dominant to the allele **b** for black body. Explain why, in the offspring of a mating between a pure-breeding black female and a yellow male, all the males will be black.   [1]

   Male offspring all contain a single b only ✔. ◄

*Take care with a question like this! Males are produced from an unfertilised egg. So the parent is female and **bb**. All haploid males must be **b**. There is no fusion of gametes to produce a male.*

[AQA A Specimen]

A grade candidate – mark scored 12/13

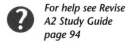

For help see Revise A2 Study Guide page 94

(2) The diagram below shows a pedigree for fibrocystic disease.

Examiner's Commentary

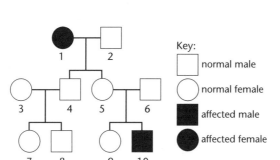

Key:
☐ normal male
○ normal female
■ affected male
● affected female

(a) (i) Using the above diagram, give **two** reasons for concluding that the allele for fibrocystic disease is recessive. **[2]**

1. The disease skips a generation ✔.
2. Parents 5 and 6 do not have cystic fibrosis but their son does ✔.

If the condition was dominant parents 5 and 6 would have the condition. As it is, they produce 10 who has the bb (homozygous) condition.

(ii) Explain how you could conclude from the above information that the allele for fibrocystic disease is not located on either the X or the Y chromosome. **[2]**

X chromosome - If the condition was X linked then male 4 would have inherited the condition from female 1 ✔.
Y chromosome - the condition is not confined to males ✔.

The candidate realised that female 1 had two cystic alleles. If sex linked, on the X, then Y would have had the cystic condition.

(iii) Which individuals in the pedigree are definitely heterozygous? **[1]**

4, 5 and 6 ✔

*Parent 1 is homozygous recessive and must pass on a cystic allele in **every** gamete. Parent 2 can pass on a normal (dominant) allele. 4 and 5 must be heterozygous. Parents 5 and 6 produce 10, so both 5 and 6 must be heterozygous.*

(iv) Since individuals 5 and 6 have one affected child they decided to seek the advice of a genetic counsellor before having further children. What is the chance of their next child being affected?

Explain your answer. **[1]**

1 in 4, because both of the parents are heterozygous ✔.

Correct!

Questions with model answers

A grade candidate continued

For help see Revise A2 Study Guide page 94

Examiner's Commentary

(v) In the past, people born with fibrocystic disease usually died before reaching adulthood. Suggest an explanation for the fibrocystic disease allele remaining at a relatively stable level in the population in spite of this. **[2]**

The cystic fibrosis allele is masked in the heterozygous condition, because the phenotype appears normal ✔.

.... ✗

The candidate failed to score the second mark. This was available for the fact that mutation from the normal allele to cystic allele can increase the frequency of the cystic fibrosis allele.

(b) The appropriate proportion of people suffering from fibrocystic disease is 0.04%. Use the Hardy-Weinberg equation to calculate:

- the allele frequencies in the population;

- the proportion of the population that are homozygous dominant and the proportion that are heterozygous;

- the number of people expected to be carriers of the disease in a population of 1000.

Show your working and write your answers in the spaces below. **[5]**

The fibrocystic allele frequency
Hardy-Weinberg equation $p^2 + 2pq + q^2 = 1$
$q^2 = 0.04$ so $q = $ square root of $0.04 = 0.2$ ✔
The normal (wild type) allele frequency
$p + q = 1$ therefore $p + 0.2 = 1$
$p = 1 - 0.2 = 0.8$ p ✔
The proportion that are homozygous dominant
$p = 0.8$
so $p^2 = (0.8)^2$
$= 0.64$ ✔
The proportion that are heterozygous
$2pq = 2 \times 0.8 \times 0.2$
$= 0.32$ ✔
The number expected to be carriers in a population of 1000
These are the heterozygous individuals
$0.32 \times 1000 = 320$ ✔

Always show your working. This is a typical Hardy-Weinberg question where you are given the frequency of homozygous recessives. Follow the working of the A candidate through. If you get a similar question in your live examination the principle will be the same.

[NICCEA Specimen]

Exam practice questions

Answers on p. 64

(1) In *Primula sinensis* two unlinked gene loci are responsible for the main variation in flower colour. The dominant allele **K** results in the production of a red pigment, whilst the recessive allele **k** results in the production of a pink pigment. The dominant allele **B** produces a co-pigment, whereas the recessive allele **b** results in no such production. The co-pigment forms complexes with the pigments which are bluer in colour than the pigments alone.

A homozygous pink-flowered plant, **kkbb**, was crossed with a homozygous blue-red-flowered plant, **KKBB**, and the resulting F_1 generation interbred to produce an F_2 generation.

Draw a genetic diagram in the space below to show the genotypes and phenotypes of the F_1 and F_2 generations of this cross. (Take care that the symbols **K** and **k** cannot be confused in your answer.) **[10]**

[OCR Specimen]

(2) Rats and mice are common pests. Warfarin was developed as a poison to control rats and was very effective when it was first used in 1950.

Resistance to warfarin was first reported in British rats in 1958 and is now extremely common. Warfarin resistance in rats is determined by a single gene with two alleles W^s and W^r. Rats with the genotypes listed below have the characteristics shown.

$W^s W^s$ Normal rats susceptible to warfarin.

$W^s W^r$ Rats resistant to warfarin needing slightly more vitamin K than usual for full health.

$W^r W^r$ Rats resistant to warfarin but requiring very large amounts of vitamin K. They rarely survive.

(a) Explain why:

    (i) there was a very high frequency of W^s alleles in the British population of wild rats before 1950. **[1]**

    (ii) the frequency of W^r alleles in the wild rat population rose rapidly from 1958. **[2]**

(b) Explain what would be likely to happen to the frequency of W^r alleles if warfarin were no longer used. **[2]**

(c) In humans, deaths from conditions where blood clots form inside blood vessels occurs frequently in adults over the age of fifty. These condition may be treated successfully with warfarin. However, some people possess a dominant allele which gives resistance to warfarin. Why would you not expect this allele to change in frequency? **[2]**

[AQA B Specimen]

Answers

(1)

	pink		blue-red
	kkbb	×	**KKBB**
gametes	**kb**		**KB**;
F_1 generation			**KkBb**; (genotype)
			blue-red; (phenotype)
$F_1 \times F_1$	**KkBb**	×	**KkBb**;

gametes for each **KB Kb kB kb**;

F_2

	KB	Kb	kB	kb
KB	KKBB	KKBb	KkBB	KkBb
Kb	KKbb	KKbb	KkBb	Kkbb
kB	KkBB	KkBb	kkBB	kkBb
kb	KkBb	Kkbb	kkBb	kkbb

(One mark off for each of the first two mistakes in the genotypes.)
F_2 correct genotypes and phenotypes identified

K-B- = blue-red;   kkB- = blue-pink;
K-bb = red;   kkbb = pink;

(These can also be credited in the punnet square, as above correct ratio, 9:3:3:1)

(Any 10)

(2) (a) (i) Rats with W^r allele need more vitamin K so they have a disadvantage.

Examiner's tip

Look at this answer carefully. Rats with W^r allele may not receive vitamin K so rats with W^s allele increase **proportionally**.

   (ii) The homozygous W^sW^s, were selected against,
       W^sW^r were more likely to survive and pass on the allele.   (Any 2)

Examiner's tip

Note the term 'selected against'. It means here that some of the rats, W^sW^s, die.

(b)   There would be a fall in frequency,
       W^sW^r no longer has an advantage/W^sW^s now has an advantage.

(c)   Deaths occur mainly in people over 50.
       Many people of this age have already had children.

Examiner's tip

If the people die after their mainstream reproductive age then the alleles will have already been passed on.

Questions with model answers

C grade candidate – mark scored 4/5

 For help see Revise A2 Study Guide page 100

(1) The diagram below shows the way in which four species of monkey are classified.

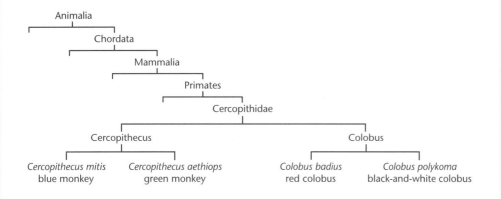

(a) This system of classification is described as hierarchical.

Explain what is meant by hierarchical. **[1]**

Beginning with the kingdom Animalia each group beneath is a smaller group ✔.

Hierarchical means that large groups are sub-divided into smaller ones.

(b) **(i)** To which genus does the green monkey belong? **[1]**

Cercopithecus ✔.

The genus is the group **before** the species. The species of green monkey is **aethiops**.

(ii) To which family does the red colobus belong? **[1]**

Cercopithidae ✔.

(c) What does the information in the diagram suggest about the similarities and the differences in the genes in these four species of monkey? **[2]**

All four species had the same group as an ancestor, Cercopithidae ✗.

Blue and green monkeys are in the same genus so closely related, whereas red colobus and black and white colobus share a different genus so are related to each other more than to the blue and green monkeys ✔.

The candidate showed understanding, but in the first part, needed to state that Cercopithidae was the **same family**.

[AQA B Specimen]

Questions with model answers

A grade candidate – mark scored 16/18

 For help see Revise A2 Study Guide page 106

(2) When breeding cattle for increased milk yield, the sequence of events shown below may be used.

pedigree bull
↓
sperm samples are collected and stored
↓
pedigree cow with high milk yield is induced to superovulate
↓
artificial insemination
↓
embryos are collected
↓
embryos are implanted into surrogate cows
↓
high milk yield female offspring

(a) (i) Describe how samples of sperm are stored. **[2]**

The semen is stored in liquid nitrogen at a low temperature ✔*.*
Straws are used after the semen has been diluted ✔*.*

A further mark was available for referring to the medium in which the semen is diluted and stored. The candidate had already gained a maximum.

(ii) Discuss the advantages and disadvantages of the use of artificial insemination in animals. *(In this question, 1 mark is available for the quality of written communication)* **[8]**

Advantages:
Only a superior bull is used, having been checked out in a breeding programme first ✔*. The farmer does not need to keep a bull which would be expensive to feed* ✔*. The semen straws are easy to transport in a liquid nitrogen transporter* ✔*.*
Disadvantages:
The very low temperatures may damage some sperms ✔*. If the same bull becomes too popular and is used too much, variation is reduced* ✔*.* ✔

Two marks were missed here. A disadvantage which could have been credited was that the artificial insemination system is expensive. Having given an answer based on expense i.e. no need to keep a bull, the expense of the artificial insemination system did not naturally follow. Do not leave gaps! Always have ideas.

Communication mark.

(b) (i) Describe how the embryos are collected after artificial insemination. **[2]**

They are flushed out of the uterus ✔ *and can be stored in liquid nitrogen conditions* ✔*.*

Correct. The candidate could have given the answer, 'the embryos are frozen'.

(ii) Explain how the surrogate cows are prepared to ensure successful implantation. **[4]**

They are treated with hormones ✔*. These hormones prepare the endometrium (uterus lining)* ✔*, which must be in a suitable condition during oestrous* ✔*. It is a surgical process to transfer the embryos into the surrogate* ✔*.*

The candidate scored maximum marks. Another could be credited, 'Several embryos can be transferred to the surrogate.' Remember that the hormones used are additional oestrogen to build up the endometrium and progesterone to maintain it.

(c) Explain how the breeder knows that a bull carries genes for high milk yield. **[2]**

Check out the milk volume per year produced by the daughters ✔*.*
This is known as progeny testing ✔*.*

[OCR Specimen]

Exam practice questions

Answers on p. 68

(1) The histogram below shows the height of wheat plants in an experimental plot.

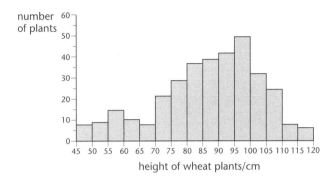

(a) What evidence from the data suggests that there are two strains of wheat growing in the experimental plot? **[1]**

(b) **(i)** Which type of variation is shown by the height of each of the strains of wheat plants? Give a reason for your answer. **[2]**

(ii) Explain why the height of the wheat plants varies between 45 cm and 120 cm. **[1]**

[AQA B Specimen]

(2) The diagram shows how four species of pig are classified.

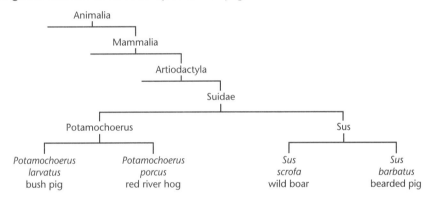

(a) **(i)** To which family does the red river hog belong? **[1]**

(ii) To which genus does the bearded pig belong? **[1]**

(b) Some biologists think bush pigs and red river hogs belong to the same species. The list below summarises some features of the biology of bush pigs and red river hogs.

- The bush pig has a body length of 100–175 cm and a mass of 45–150 kg. The red river hog has a body length of 100–145 cm and a mass of 45–115 kg.
- The red river hog is found in West Africa. The bush pig is found in East Africa.
- Both animals are omnivorous but feed mainly on a variety of underground roots and tubers.
- The ranges of these animals overlap in Uganda. In this area populations of animals which have all of the characteristics intermediate between those of bush pigs and red river hogs have existed for many years.

Do you think that bush pigs and red river hogs belong to the same or to different species? Explain how the information above supports your answer. **[3]**

[AQA A Specimen]

Answers

(1) (a) There are two peaks evident in the graph.

(b) (i) Continuous variation.
There are large number of groups.

Examiner's tip

No height categories were missing throughout the height range of wheat which pointed to continuous variation.

(ii) Phenotype is affected by both the environment and the genotype/inheritance is polygenic.

Examiner's tip

It would not have been enough to attribute just environment as the cause of height variation. It was important to link environment to its effect on genotype.

(2) (a) (i) Suidae

(ii) Sus

Examiner's tip

*Always remember 'KING PENGUINS CLIMB OVER FROZEN GRASSY SLOPES',
i.e. Kingdom, Phylum, Class, Order, Family, Genus, Species. This helps you to arrive at
Suidae (family) and genus (Sus).*

(b) There is considerable variation in body length/there is variation in the mass among members of a species,
Many different species have broadly similar diets.
There are intermediates between the bush pigs and red river hogs suggesting that they can interbreed.
These intermediates or hybrids are capable of breeding successfully, 'for many years'.

(Any 3)

Examiner's tip

It was important to decide that they belong to the same species. This was confirmed by the final two bullet points. Organisms from the same species can interbreed and produce fertile offspring.

Questions with model answers

C grade candidate – mark scored 4/6

 For help see Revise A2 Study Guide page 110

Examiner's Commentary

(1)   The wren is a small insect-eating bird. The percentage change in the size of the wren population from one year to the next was estimated over a number of years. The number of days with snow lying in the previous winter was also recorded.

This information is shown on the graph below.

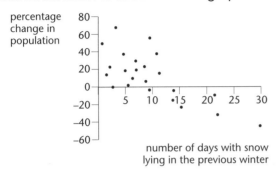

(a)   (i)   Describe the relationship between the number of days with snow lying and the change in population size.   **[2]**

The number of wrens in the population decreases when snow covers the ground for a greater number of days ✔. *More wrens survive when the snow covers the ground for fewer days* ✗.

The second point was merely a repeat of the first! The Examination Board awarded the second mark for the fact that the population decreased after 12 days.

(ii)   Suggest and explain a reason for this relationship.   **[2]**

The food is probably under the snow, so the wrens cannot reach it ✔ *so some of the birds die* ✗.

The second mark could not be awarded. The Examination Board awarded this mark for the fact that there were fewer wrens to breed. Less birds lay less eggs so the population may fall.

(b)   A comparison was made between the number of breeding pairs of wrens each year and their breeding success.

Number of breeding pairs of wrens /millions	Percentage increase in population size
1.2	55
1.9	48
2.5	35
2.9	25
3.9	2

Suggest an explanation for the relationship between the size of the breeding population and breeding success.   **[2]**

The more breeding pairs there were, because of competition for food, the less the wren population increased in percentage terms ✔. *After breeding the numbers surviving decreased* ✔.

The candidate analysed the table correctly. The competition which takes place is intraspecific because of the same species sharing resources.

[AQA Specimen]

Questions with model answers

A grade candidate – mark scored 14/15

 For help see Revise A2 Study Guide page 110

Examiner's Commentary

(2) In this question you are expected to answer in continuous prose, supported, where appropriate, by diagrams. You are reminded that up to 2 marks are awarded for the quality of written communication.

Sand dunes may differ in their species composition. The diagram below is a vertical section through one particular sand dune system

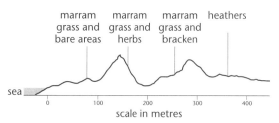

scale in metres

(a) Describe how you would sample the area depicted in the diagram in order to illustrate quantitatively the zonation of the vegetation. **[6]**

I would use a belt transect ✔. The belt transect would be taken from the position I have shown on the diagram in this question ✔. The transect would be made up of a line of quadrats ✔, and in each one I would identify each plant species ✔. In every quadrat I would estimate the percentage cover or measure the density of each of each plant species ✔. I would collect the data and write it into a table ✗.

*The answer is strong but the final mark could not be awarded. The Examination Board required that a table had **appropriate headings**, i.e. quadrat position, or percentage cover/ density per quadrat of each species. The candidate showed the direction of the belt transect on the diagram. This was a good technique!*

(b) Ground beetles are also found in the area. They are most likely scavengers and are particularly active during the hours of darkness. It is suggested that there are two separate species of ground beetle in the different areas of the sand dunes. How would you investigate this suggestion? Describe the procedure as fully as possible. **[7]**

I would use a pitfall trap like this:

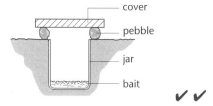

✔ ✔

The pitfall traps would be placed across each dune area ✔. I would put appropriate bait into each trap to attract scavengers ✔. I would use a large number of traps in each area ✔ then take an average for each type of ground beetle captured ✔.
Classification of the beetles would be important so I would use an identification key ✔.
I would write the data for each type of ground beetle into a table.

✔✔ (Communication)

[NICCEA Specimen]

Full marks! The candidate described a suitable method of trapping the scavengers and used the option of drawing a labelled diagram. Communication marks were available. Here are the criteria which were satisfied by the answer. Communicates clearly and coherently, through well-linked sentences and paragraphs. Arguments are generally relevant and well sequenced. When appropriate, biological terminology is used accurately. There are few errors of grammar, punctuation and spelling.

Exam practice questions

Answers on p. 73

(1) The diagram shows a vertical section through an area of tropical rainforest in Malaysia.

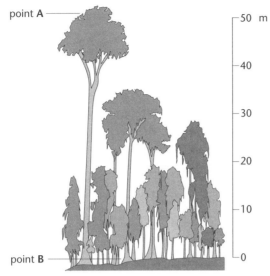

The graph shows the daily fluctuations in carbon dioxide concentration at points A and B.

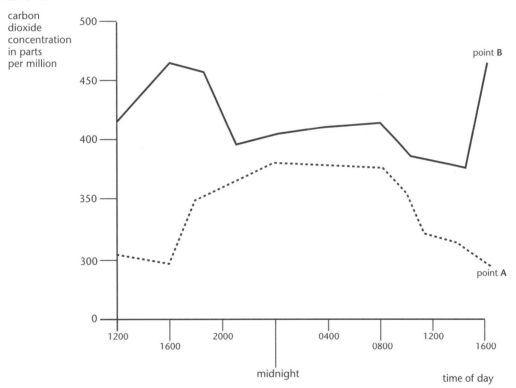

(a) (i) Explain the fluctuations in the carbon dioxide concentration which take place at point A over a 24-hour period. [1]

(ii) At 1600 hours, the carbon dioxide concentration at points A and B differ from each other. Explain why. [2]

(b) Use information in the diagram to help explain why replacing rain forest with agricultural crops will reduce the diversity of animals. [2]

[AQA A Specimen]

Exam practice questions

(2) Shell colour and pattern in *Cepaea nemoralis*, the land snail, is genetically determined. Shells may be brown, green or pink. In addition, the whole shell may have up to five dark bands on it. Thrushes eat these snails, which they select by sight.

The table shows the distribution of some shell types in three different habitats.

Shell type	Percentage of each shell type present in:		
	Grassland (green pasture)	Rough herbage (nettles, long grass and dead stems)	Woodland floor (dead leaves, soil and twigs)
Brown	10	23	54
Green	58	19	12
Pink	7	8	4
Banded	25	50	30

(a) (i) Name the shell type with the greatest survival value on the woodland floor. [1]

(ii) Explain the relative abundance in grassland and rough herbage of snails with:
1. green shells, [1]
2. pink shells. [1]

(iii) Explain **fully** the relative abundance of banded snails in grassland, rough herbage and woodland. [2]

(b) During very hot summers pink shells and pale coloured shells are more frequent.

(i) Suggest **one** factor, apart from predation, which might influence selection under these conditions. [1]

(ii) Suggest **one** possible selective advantage, to the snails, of pale coloured shells. [1]

[WJEC Specimen]

Answers

(1) (a) (i) The carbon dioxide falls in the day time because it is used in photosynthesis and increases at night when photosynthesis ceases.

(ii) At B there is very little light which limits photosynthesis,
Respiration of animals produces carbon dioxide/the decomposers produce carbon dioxide.

Examiner's tip

Always remember that as CO_2 falls it may be used by plants for photosynthesis. Part (ii) is more difficult. In response to a graph question like this do not read off just by 'eye'. Use a ruler, here, to draw a vertical line up from 1600 hours, then horizontal lines to the x axis. The CO_2 'read off' becomes more accurate.

(b) Agricultural crops have fewer species/crops may have fewer layers of vegetation.
There is a smaller range of food/there are fewer niches available.

Examiner's tip

It would be necessary to destroy habitats to make way for the crops. It is logical to suggest that specific foods or homes offered by the forest would no longer be available. Thr canopy of the forest offers several different niches.

(2) (a) (i) Brown shell

(ii) 1. Green shells are better camouflaged/they are less conspicuous in the green grass/they stand out against the mixture of green and brown stems/or the leaves of the rough herbage.
2. The pink shelled snails are most visible to predators in both habitats.

Examiner's tip

The key fact here is the colour of shell of each type of snail. When giving the answer 'camouflage' it is always good practice to state 'camouflage from predators'.

(iii) The banded shell pattern breaks up the shell shape/the banded pattern gives a disruptive/mixed colour image.
Rough herbage gives a mottled background on which the banded snail is camouflaged.
Dark bands form a contrast against the uniform green pasture/light bands form a contrast against the uniform brown of the woodland floor. (Any 2)

Examiner's tip

*The question pointed to a full explanation so that detail is the key. The relative abundance figures in the table showed which snails were ideally camouflaged and the ones which were not, in **each** habitat.*

(b) (i) Temperature

(ii) The pink shell reflects heat/it may provide a physiological advantage.

Examiner's tip

In a 'suggest' question it is usual that there are several alternatives. Be prepared to give logical ideas in your responses.

Questions with model answers

C grade candidate – mark scored 4/6

For help see Revise A2 Study Guide pages 122 and 119

(1)   Describe concisely the cause and the consequences of each of the following.

(a) Acid rain   **[3]**

> *Sulphur dioxide is produced as a result of the burning of fossil fuels. It enters the air and finally dissolves into the water vapour of the clouds to form acid* ✔. *As a result of the toxic effect of acid rain the plants are affected* ✔✗.

(b) Ozone depletion   **[3]**

> *CFCs cause the hole in the ozone layer. They are produced as a result of some aerosols* ✔. *The ozone layer, when complete, reduces some of the ultraviolet light which reaches the surface of the Earth* ✔. *More ultraviolet light reaches the surface which has a bad effect on people.*

[NICCEA Specimen]

Examiner's Commentary

The candidate could have given a nitrogen oxide instead of sulphur dioxide. It is also correct. The toxic nature of the acid rain could also be credited, but a mark was missed by the statement that the plant life was affected. This is much too vague! It defoliates trees and results in their death. Giving death of trees would have gained the final mark. Additionally fish death due to aluminium release clogging up the gills would have scored an alternative two marks. Similarly the acid cause of bronchitis would score two marks.

The candidate began well but again gave a vague answer. Compare the following answer with the final part of the one given. 'Ultraviolet light increases the incidence of skin cancer.' Better use of technical words, in context, will be more likely to be awarded credit.

A grade candidate – mark scored 16/19

(2)   The figure shows a section of a river that flows through rural and urban areas.

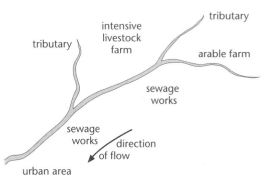

(a) Explain how the biological oxygen demand (BOD) of water samples taken from the river depicted in the figure could be measured.   **[4]**

> *Firstly I would take equal samples of water from a number of positions along the river* ✔. *I would find the oxygen level of the water at the beginning, in each sample* ✔. *Each sample of water would be kept in darkness, at a temperature of 20 °C for five days* ✔✔.
>
> *The BOD is the volume of oxygen used by the bacteria in the water to decompose the organic matter contained in it.*

The candidate scored the maximum 4 marks then went on to give two more points which are worthy of credit. You could not answer this question without recalling the BOD technique. The Examination Board also gave credit for an alternative method; add a small drop of methylene blue; so does not mix with oxygen; at 20 °C; time how long it takes to go colourless.

A grade candidate continued

For help see Revise A2 Study Guide page 119

(b) Outline **three other** ways in which the quality of the water in the river may be assessed. **[5]**

> 1. The suspended solids could be measured ✔. This could have been found by taking a sample of water and filtering out the particles. After drying the mass of solids could be measured ✔.
>
> 2. The oxygen concentration of the water could be measured ✔. You could do this like we did at our school. We used a probe linked to an interface and datalogger.
>
> 3. The cloudiness of the water could have been measured ✘. I would use a colorimeter for this ✔.

(c) Explain the sequence of events that may lead to the eutrophication of the river shown in the figure. **[9]**

(In this question, 1 mark is available for the quality of written communication.)

> Fertilisers containing nitrates are used on crops. Sometimes as soon as they are put on the crops it rains, so they run off the surface of fields into the river ✔ ✔. Also if sewage or silage effluent enter the river, they also result in nitrate increase in the river water ✔. The effect of the nitrate is to promote the growth of the plant life of the river ✔.
> The effect is that algae near the surface of the river grow in numbers forming an opaque carpet over the surface ✔. Light entry for bottom dwelling plants is prevented.
>
> If raw sewage reaches the river then the eutrophication effect is increased ✔. The bacteria in the water decompose the sewage, releasing the minerals which boost eutrophication ✔.
>
> ✔

[OCR Specimen]

Examiner's Commentary

The candidate answered the question well but in the third part stated 'cloudiness'. The correct term is <u>turbidity</u>.
Other answers could have been given:
assessing bacterial numbers by Petri dish techniques or direct counts,
assessing the salinity of the water by use of a conductivity meter,
assessing pH using a pH meter.
It is necessary to point out a danger in answering this type of question. The computer and probe technique could have been used for pH, oxygen and salinity. Many Examination Boards would credit the use of an environmental probe only once. Vary your techniques!

The candidate could be awarded the communication mark. It was borderline because this Examination Board requires the use of specialist terms, with clarity, in an organised way.

Exam practice questions

 Answers on p. 77

(1) A diversity index can be used to compare the level of pollution at different places along a stream. It may then be calculated from the number of small animals collected in the stream. These animals cling to plants or stones or live in the bed of the stream.

    (a) (i) What information is needed to calculate a diversity index? **[2]**

        (ii) Describe **one** technique which could be used to capture the small animals in order to calculate the index. **[2]**

        (iii) Give **two** precautions which must be taken in order to make a valid comparison of diversity of organisms at different places along a stream. **[2]**

    (b) Suggest **one** factor that could affect the concentration of oxygen in a stream and explain how it would have its effect. **[2]**

[AQA B Specimen]

(2) Most cereal fields in Britain are sprayed with selective herbicides. In order to conserve wildlife, farmers are recommended to leave unsprayed a 6 metre strip, called a headland, around each cereal field.

    (a) Explain how spraying selective herbicides on a headland might affect the number of insects living there. **[2]**

    (b) The table below shows the results of an investigation to find the effect of leaving headlands unsprayed on the population of butterflies living there.

Butterfly species	Number of each species recorded on headland sprayed with selected herbicide	Number of each species recorded on headland not sprayed with selective herbicide	χ^2	Significance (NS = not significant)
Small skipper	2	41	35.4	P < 0.001
Large skipper	1	17	14.2	P < 0.001
Large white	38	56	3.4	NS
Holly blue	13	29	6.1	P < 0.05
Hedge brown	59	93	7.6	P < 0.05
Small heath	0	11	11.0	P < 0.01
Ringlet	23	52	11.2	P < 0.01

        (i) Why was a χ^2 (chi-squared) test applied to the results? **[1]**
        (ii) What conclusions can be drawn from the results of the χ^2 test? **[3]**

[AQA B Specimen]

Answers

(1) **(a)** **(i)** The total number of organisms of all species is needed.
The total number of organisms of each species is needed.

Examiner's tip

On this occasion you did not need to actually calculate a diversity index. Often the equation is given and you would need to substitute the numbers identified above, in the equation.

(ii) A net could be swept through the stream to catch the animals.
The net to be used systematically from same size of area for each sweep, i.e. standard area method.

Examiner's tip

The Examination Board would credit a good idea for capture. The sweep net may be used in the same way each time through the same volume and depth of water.

(iii) Random selection of sampling areas at each site along the river.
Must test a large number of samples, to make sure the results are statistically valid.

Examiner's tip

*The aim was to capture animals in order to calculate the diversity index. The key term to use is **random** because many areas of the river need to be sampled, e.g. middle, edge, deep, shallow, up river, down river.*

(b) Factor – shallow area/stony area.
Explanation – this gives the oxygen more chance to dissolve into the water.

(2) **(a)** The herbicides kill the weed which may be the food plants of some of the insects which live there, so numbers of these insects decrease.
The insects which died may, themselves, be the food source for other insects.

Examiner's tip

Herbicides are weedkillers and do not kill insects outright! Awareness of the consequence of plants dying was needed here. In environmental type questions it is useful to think in terms of food chains and webs.

(b) **(i)** To accept or reject a null hypothesis/to see if differences were significant.

Examiner's tip

It is not enough to claim that your experimental results are significant. You have to prove it. Chi-squared helps you to do just this. The test shows the probability of results being significant or due to chance alone.

(ii) Spraying had no effect on the population of large white butterflies.
Spraying almost certainly affected the population of small skipper butterflies/less than 1 in 1000 probability that the reduction was due to some other factor.
There is a greater chance that reduction in population of other butterflies is by chance/less than 1 in 100 or 1 in 500.

Examiner's tip

*Look carefully at the significance column in the table. $P < 0.001$ means less than 1 in 1000 probability that the reduction in butterflies was due to some other factor. In other words, the results **are significant!***

Questions with model answers

C grade candidate – mark scored 3/5

For help see Revise
A2 Study Guide
pages 129 and 126

Examiner's Commentary

(1)   Read the following passage on gram staining and then write on the dotted lines the most appropriate word or words to complete the account.

Gram positive and Gram negative bacteria differ in their ability to keep a stain in their cell walls. A slide of bacteria is first flooded with a suitable stain such as *crystal violet* ✔. All bacteria absorb this stain, which is then fixed with Gram's *iodine* ✔ before being decolourised with *distilled water* ✗.

Gram positive bacteria retain the stain complex, but Gram negative do not. Counterstaining with red stain, such as *safranin* ✔ causes Gram negative bacteria to absorb it and become red. Gram positive bacteria remain *red* ✗ in colour.   [5]

*This cloze exercise was based on the gram's staining technique. The information in an unfinished passage should aid your recall. The C Candidate was incorrect in two places. Gram's **iodine** is used to fix, (together with heat). Gram positive bacteria remain **purple** in colour.*

A grade candidate – mark scored 20/21

(2)   The figure shows a labelled diagram of a bacteriophage.

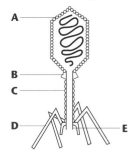

(a) Name the group of microorganisms which includes bacteriophages.   *Viruses* ✔   [1]

Remember that bacteriophages are viruses which attack bacteria.

(b) Using the letters on the figure, identify the region:

    (i)   involved in the attachment to bacteria.   *D* ✔   [1]

    (ii)   involved in penetrating bacterial cells.   *E* ✔   [1]

    (iii) containing genetic information.   *A* ✔   [1]

(c) Describe the life cycle of a **named** bacteriophage. *(In this question, 1 mark is available for the quality of written communication.)*   [9]

The bacteriophage T_2 ✔ is a virus which attacks the bacterium E. coli. The T_2 bacteriophage makes contact with the outer wall of the bacterium. Part D attaches to the cell wall of the bacterium ✔. The bottom of the tail sheath at E is the tail plate. When the tail plate contacts the cell wall an enzyme is produced which makes a small hole in the cell wall ✔. The tail sheath contracts injecting the viral DNA into the bacterium. Once the DNA is inserted into the bacterium there are two possible consequences,

A grade candidate continued

 For help see Revise
A2 Study Guide
pages 126 and 129

Examiner's Commentary

1. The viral DNA can become incorporated into the host DNA. Without
destroying the bacterium it replicates in each future generation of bacteria.
This is called lysogeny ✔*.*
2. The viral DNA can become incorporated into the bacterial DNA The
viral DNA takes over the protein synthesis of the bacterium, making viral
proteins ✔*. This results in viral enzymes being produced which are able to*
synthesize capsomere proteins for the head of the virus ✔ ✔*. New viruses*
are produced within the bacterium which finally escape as the bacterial
membrane and wall split. This is known as lysis ✔*.*
(Communication mark) ✔

[Edexcel Specimen]

> *All correct. The A*
> *Candidate gave all*
> *required detail. There were*
> *more marks available up*
> *to the maximum of 8 for*
> *the main answer. The*
> *candidate failed to state*
> *that viral DNA is made*
> *during the attack.*

The number of bacteriophages in a liquid medium can be
determined by serial dilution of the medium, followed by plating
a small volume, 0.5 cm^3, on to an agar plate that has been
covered in bacteria. Each bacteriophage is capable of infecting a
bacterium. When the agar plates are incubated, a bacterial lawn
results in which there are clear areas known as plaques. A
plaque is produced when a single bacteriophage infects a
bacterium, and eventually results in the death of a large number
of bacteria. The clear areas are due to lysis of these bacteria.

The number of plaques produced from a serial dilution of the
medium on two sets of plates, inoculated with the same
bacterium, are shown in the table.

	dilution		
	10^{-6}	10^{-7}	10^{-8}
number of plaques	956 948	98 94	7 3

(d) With reference to the table:

   (i) suggest why the values at the dilutions of 10^{-6} and 10^{-8} are
inaccurate and should not be used to estimate
bacteriophage numbers. **[5]**

The 10^{-6} plates have plaques so great in number that they will overlap
each other. This is not accurate for counting them ✔ ✔*.*
The 10^{-8} plates have very few plaques ✔*. This would result in a greater*
chance of statistical errors ✔*.*

> *The candidate answered*
> *well but missed a potential*
> *mark. The 10^{-8} plates*
> *were made using very*
> *dilute medium. Perhaps, if*
> *it was not stirred correctly,*
> *there could be significant*
> *dilution errors.*

   (ii) Estimate the number of bacteriophage particles per cm^3 in
the original liquid medium. Show your working. **[3]**

$$\frac{94 + 98}{2} \; ✔ = 96$$

$$\frac{1\,cm^3 \times 96 \times 2}{10^{-7}} \; ✔ \quad = \quad 1.92 \times 10^9 \; ✔$$

[OCR Specimen]

Exam practice questions

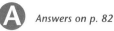

Answers on p. 82

(1) The graph summarises the main events which take place during the production of the antibiotic penicillin by the fungus *Penicillium notatum*.

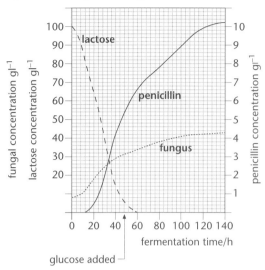

(a) (i) Describe **three** differences between the pattern of fungal growth and the pattern of penicillin production. **[6]**

(ii) Calculate the rate of penicillin production per hour during the period 32 to 40 hours. Show your working. **[4]**

(b) (i) Penicillin production was maintained by adding glucose after 48 hours. Explain why this **particular** source of energy for ATP was added instead of more lactose. glucose - immediate source Lactose need be br[1]dn **[1]**

(ii) Suggest **two** conditions which should be controlled in order to achieve the maximum rate of penicillin production. Temp. P*, O₂ Supply **[2]**

(c) (i) Before cheeses are stored they are often covered with a coating containing an antibiotic. Suggest the purpose of this procedure. Prevents growth [1] ng! **[1]**

(ii) Suggest why penicillin would not be chosen as the antibiotic in this coating. - allergic 2 penculin - no Its 4 humans **[1]**

[WJEC Specimen]

(2) The sequence below shows some essential steps in the manufacture of yoghurt from milk.

A Homogenisation of the milk (breaks up large fat globules)
B Pasteurisation (heated to 72 °C for 10 seconds)
C Fermentation (starter culture added)
D Stirring and cooling
E Addition of flavouring and colourings
F Packaging

(a) Why is pasteurisation of milk carried out at step B? To kill bacteria. **[1]**

(b) At step C, a starter culture is added to bring about the fermentation. During this stage lactose in the milk is broken down, resulting in the formation of lactic acid. Two organisms, P and Q, are commonly used in starter cultures. Each one on its own is capable of bringing about fermentation, but may be used together.

Lactobacillus
Lactobugin

The graph below shows the rate of lactic acid production when the two organisms are used separately (single strain cultures) and when they are used together (mixed strain cultures).

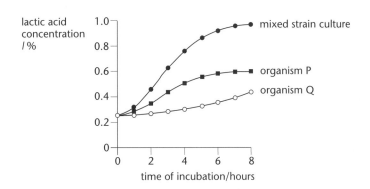

Name the two organisms commonly used in starter cultures. **[2]**

(c) Suggest which type of starter culture is the most suitable, giving a
reason for your answer. *Mixed strain, faster rate of production* **[2]** *p H*
(d) Explain the importance of lactic acid in the production of yoghurt. — *redus* **[2]** *P +*
(e) State why, at step D, the yoghurt is cooled rapidly to 5 °C. **[1]** *+ flavour*
   Inhibit activity bacteria

[Edexcel Specimen]

Answers

(1) (a) (i) 1. In the lag phase or from 0 to 12 hours the fungal cells are becoming accustomed to the conditions/at this time they reproduce very slowly, so no penicillin is produced.
2. Fungal cell increase is greatest from 10 to 30 hours,
penicillin or antibiotic increase is greatest from 22 to 47 hours.
3. Fungal cell population levels off after 48 hours, penicillin amount continues to increase.
(fungal cell reproduction decreases before penicillin production reaches a peak = 2)
4. Penicillin production starts to level off at 120 hours,
This takes place around 80 hours after the fungus has begun to level off.

(Any 3 pairs)

(ii) In 8 hours (32–40 hours) the penicillin increases by 1.8 gl^{-1},
so in 1 hour 1.8/8 gl^{-1} will be produced.
Answer = 0.225 or 0.2 (acceptable)
gl^{-1}. (unit mark)

Examiner's tip

In the question the candidate was informed of the time period, 32–40 hours. These points need to be marked on the penicillin graph line and horizontal lines drawn to the x axis. The read off becomes accurate in this way! Note that marks are often given for the units. Never leave them out.

Answers

(b) (i) Glucose is an immediate source of energy, lactose must first be broken down to produce glucose.

(ii) Oxygen supply, Temperature, pH

(Any 2)

Examiner's tip

The food source is important. Penicillin must produce lactase, the enzyme, to break down the glucose. The conditions are typical of those required by many microorganisms. Even if you knew nothing about penicillium, you could guess these.

(c) (i) Prevents the cheese being spoiled/prevents growth of surface microorganisms or fungi or bacteria.

(ii) It is used in the treatment of human disease/indiscriminate use results in the production of resistant microorganisms which may be a health hazard/some people are allergic to penicillin.

(2) (a) To kill pathogenic microorganisms/to kill pathogenic bacteria.

(b) 1. *Streptococcus thermophilus/Lactococcus thermophilus*
   2. *Lactobacillis bulgaricus*

Examiner's tip

The task was to name two organisms commonly used in starter cultures. In addition to the above organisms Bifidobacterium bifidum *and* Lactobacillis acidophilus *are part of the starter culture for the 'Bio' range of yoghurts. Not part of the mark scheme of the examination board it is likely that they both would be credited. Read your yoghurt labels!*

(c) Mixed strain. They have a faster rate of lactic acid production/reference to synergism/reference to mutualism or symbiosis.

Examiner's tip

Clearly the graph shows that individually the strains of bacteria are not as good as the mixed. The faster rate of production of lactic acid is equally visible on the graph. An alternative to this was a reference to a synergistic relationship. This means that two things working together are more effective than the individuals working alone.

(d) Lactic acid reduces the pH to 4.6 or 4.7; Casein or milk protein coagulates at this stage; Yoghurt thickens; Gives a characteristic taste or flavour. (Any 2)

Examiner's tip

This answer illustrates how accurate you need to be at A2 Level. Lactic acid reduces pH would not be credited. This must be accompanied by the pH value 4.6 or 4.7.

(e) Reduce the activity of the bacteria/inhibit the activity of the bacteria/reduce the production of lactic acid/reduce the production of lactic acid.

Examiner's tip

After the bacteria have been so involved in the process it is important that their activity slows down or stops. If they were kept in warm conditions pressure would increase in the yoghurt pot. The top could be blown off!

Mock Exam

Time: 1 hour 10 minutes   Maximum marks: 57

Section A

(1)   The diagram below shows a section of a human heart at a specific stage in the cardiac cycle.

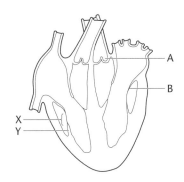

(a)   Name the parts labelled A and B.

A...

B.. **[2]**

(b)   Name the stage of the cardiac cycle shown in the diagram and give TWO reasons for your choice.

Name of stage..

Reason 1 ....

Reason 2 .... **[3]**

(c)   Give ONE function of each of the parts X and Y.

X..

Y.. **[2]**

[7 marks]

[Edexcel Specimen]

(2) Sucrose is a disaccharide. It is also a non-reducing sugar.

(a) Benedict's solution is used to identify non-reducing sugars.

Complete the flow chart to show the steps you would carry out in order to get a positive result.

Add (i) ....

⇩

(ii) ....

⇩

Add Benedict's solution then heat in a water bath

⇩

Red-brown precipitate indicates a non-reducing sugar

[2]

(b) A sample of sucrose was incubated with the enzyme sucrase. This hydrolysed the sucrose. A chromatogram was produced by loading chromatography paper with sucrose, with sucrose which had been incubated with sucrase, and with sucrase. The diagram below shows this chromatogram.

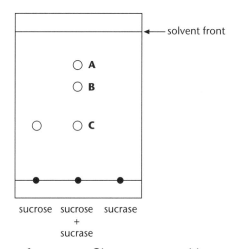

(i) Calculate the R_f value of sucrose. Show your working.

R_f value = ...   **[2]**

(ii) Explain the presence of the three spots labelled **A**, **B** and **C**.

....

....

....

[2]

[6 marks]

[AQA B Specimen]

(3) The figure shows the flow of energy through the trees in a forest ecosystem. The numbers represent inputs and outputs of energy in kilojoules per m^2 per year.

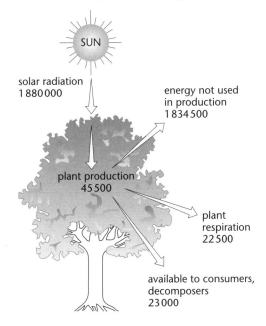

SUN

solar radiation
1 880 000

energy not used
in production
1 834 500

plant production
45 500

plant
respiration
22 500

available to consumers,
decomposers
23 000

(a) **(i)** On the figure, draw a ring around the number which indicates the energy entering the system via photosynthesis. **[1]**

(ii) The total energy available to the plant in the ecosystem is 1 880 000 kj per m^2 per year.

Calculate the efficiency of photosynthesis. Show your working.

Efficiency = ... **[2]**

(b) Suggest **four** reasons why so much solar energy is **not** used in the production in the forest ecosystem.

1. ....

2. ....

3. ....

4. .... **[4]**

(c) In what form will energy from plant respiration escape from the ecosystem?

.... **[1]**

[8 marks]

[OCR Specimen]

(4) An experiment was carried out to determine what happens to amino acids after they are absorbed by animal cells. The cells were incubated for 5 minutes in a medium containing radioactively labelled amino acids. The radioactive amino acids were then washed off and the cells were incubated in a medium containing non-radioactive amino acids.

Samples of the cells were taken at 5, 10 and 45 minutes after the start of the experiment and the sites of the radioactivity in the cells were determined.

The results are given in the table below. The figures show radioactivity in certain cell organelles expressed as a percentage of the total radioactivity within the cells.

Organelle	Percentage of total radioactivity		
	At 5 minutes	At 10 minutes	At 45 minutes
Rough endoplasmic reticulum	80	10	5
Golgi aparatus	10	80	30
Secretory vesicles	0	5	60

(a) Name ONE type of molecule synthesised from amino in cells.

.... [1]

(b) Explain why the radioactivity is associated mainly with the rough endoplasmic reticulum after the first 5 minutes of the experiment.

....

....

.... [2]

(c) Explain the changes in the pattern of radioactivity in the cell during the remaining 40 minutes of the experiment.

....

....

.... [3]

(d) Suggest why the figures in the table total less than 100%.

....

....

.... [2]

(e)   If the experiment is continued for a further period of time, most of the
radioactivity will be found outside the cell.

Name and describe the process which brings about this result.

....

....

....

.... **[3]**

[11 marks]

[Edexcel Specimen]

Section B

(1)   Read the following passage.

Forensic science has come a long way from the magnifying glass and deerstalker
image of Sherlock Holmes. Take the identification of blood stains as an example.
There has been a change in the sensitivity of tests used to distinguish between
blood obtained from different individuals. Early techniques relied on methods
5   based on the biology of red blood cells; more modern ones are based on white
blood cells.

Until relatively recently all that a forensic scientist could do was identify the blood
groups concerned. Certain protein molecules in the blood act as antigens and it is
the presence of these which determines blood group. The four blood groups A, B,
10 AB and O are determined by the presence of the relevant antigen. For example,
individuals with blood group A, have antigen A, while those with blood group B
have antigen B.

More modern techniques have allowed us to progress much further. Instead of
looking at a rather limited range of proteins we can now look at DNA itself. Genetic
15 fingerprinting can be used to distinguish between individuals by looking at
similarities and differences in part of their DNA. Some of the non-coding DNA
consists of short sequences of bases which may be repeated. The actual number
of times these sequences are repeated varies from individual to individual. Genetic
fingerprinting compares these sequences. The flow chart summarises the main
20 steps involved in the procedure.

```
┌─────────────────────────────────┐
│      DNA is extracted from a suitable  │
│            blood sample.               │
└─────────────────────────────────┘
                  ↓
┌─────────────────────────────────┐
│   Enzymes are used to cut the DNA in the │
│      sample into smaller pieces.        │
└─────────────────────────────────┘
                  ↓
┌─────────────────────────────────┐
│     The pieces of DNA are separated by   │
│   electrophoresis. This produces a sheet of │
│      gel with bands of DNA arranged on it │
│    rather like the rungs on a ladder. The │
│     smaller pieces travel farther than the │
│              larger ones.               │
└─────────────────────────────────┘
                  ↓
┌─────────────────────────────────┐
│   Particular bands of DNA are now located │
│     by using a DNA probe. The position of │
│   these bands on a gel sheet can be used to │
│           compare individuals.          │
└─────────────────────────────────┘
```

Use information from the passage and your own knowledge to answer the following questions.

(a) Explain why early techniques used to distinguish between blood obtained from different individuals relied on methods based on the biology of red cells while more modern ones are based on white blood cells (lines 4–6).

....

....

....

.... **[3]**

(b) A particular sample of blood was tested in the laboratory. It agglutinated when it was mixed with anti-A antibody; it also agglutinated when mixed with anti-B antibody.

   (i) What was the blood group of the blood sample?

.... **[1]**

   (ii) Explain how you arrived at your answer.

.... **[1]**

(c) Explain what is meant by 'non-coding' DNA (line 16)

....

.... **[1]**

(d) **(i)** Name the type of enzyme used to cut DNA in the sample into smaller pieces (box 2).

.... **[1]**

(ii) Explain why the lengths of pieces produced by cutting the DNA with one of these enzymes will vary from individual to individual.

....

....

....

....

....

.... **[3]**

(e) Describe how a DNA probe may be used to find a particular band of DNA.

....

....

....

....

....

....

....

.... **[5]**

[15 marks]

[AQA A Specimen]

(2) Answer **ONE** of the following questions.

Either – (a) Explain what is meant by translation and describe how the cell carries out translation.

Or – **(b)** Describe the **changes** that occur in a cell dividing by meiosis from the point at which the chromosomes first become visible under the light microscope.

....

....

....

....

....

....

....

....

....

....

....

....

....

....

....

....

....

....

[10]

[WJEC Specimen]

Answers

How to use the mark scheme

Symbol	Meaning
;	A separate mark
/	An alternative answer acceptable
max	Mark scheme shows that marks available exceed the question value You can be awarded up to the maximum
underline	When the word or phrase is underlined then it must be given if a mark is to be awarded

Section A

(1) (a)   A = (Aortic) semilunar valve;   B = Bicuspid/mitral/left atrioventricular valve;

Examiner's tip

This is a starter question on a paper which you should be able to answer. Success will build your confidence. Note that the full valve name is required, i.e. just valve = 0 but when qualified by semilunar the credit can be given.

(b)   The stage is (ventricular) diastole/atrial systole;
Reason 1   Aortic/pulmonary artery/semilunar valve(s) closed;
Reason 2   Bicuspid/mitral/tricuspid/atrioventricular valve(s) open;

Examiner's tip

The key to answering this question is the condition of the valves. If a valve is open then the chamber behind it is contracting (systole), whilst the chamber in front is relaxing (diastole). The reverse of this, i.e. contraction in a lower chamber, followed by relaxation in a chamber above closes a valve.

(c)   X = prevents the valve inverting;
Y = adjusts the tension in the valve/contracts to pull on the valve;

Examiner's tip

Part X represents the cordae tendonae. Ventricular walls have thicker muscle than the atria. When the ventricles contract there is a danger that the atrioventricular valves will push back into the atria. Valve collapse would result in blood backflow. The cordae tendonae prevent this.

(2) (a) (i) Hydrochloric acid;   **(ii)** Neutralise/add alkali;

Examiner's tips

Make sure in your live exam you do not forget the biochemical tests of the unit. Non-reducing sugars give a negative Benedict's test and usually would be carried out first. After a negative Benedict's test you implement the hydrolysis of the disaccharide with HCl. This breaks the glycosidic bond to release monosaccharide molecules which go on to give a positive test.
You cannot afford to get the biochemical tests wrong. Later questions will be more difficult!

(b) **(i)** 0.35

(If the sucrose spot is identified and it is recognised that R_f involves the distance moved by the spot and solvent front, it gains one mark, even if the calculation is incorrect.)

Examiner's tip

Remember the calculation. $R_f = \dfrac{distance\ moved\ by\ spot}{distance\ moved\ by\ the\ solvent\ front}$

*This is a good example of how you get partial marks for a calculation even though you give the wrong final answer. There are always marks for **showing your working**.*

(ii) Spots A and B are different sugars/monomers/named sugar produced on hydrolysis/breakdown; C is sucrose (not hydrolysed);

Examiner's tip

*Remember that the largest molecules on the chromatogram move more slowly, the smallest more quickly. In this question some of the sucrose **C** moves slowly from the start line. However, once the enzyme has acted the sucrose is broken down into two different sugars. Which sugar is the smallest?*

(3) **(a)** **(i)** A ring around 45 500; **(ii)** $\dfrac{45\ 500}{1\ 880\ 000}$; Efficiency = 2.42% *(Accept 2.4)*

Examiner's tip

The mark scheme gives a mark for the division shown. This needs to be multiplied by 100 to create the percentage 2.4.

(b) Not all light/not all solar energy/is absorbed in photosynthesis; Some energy is dissipated as heat; Some light is reflected; Some light misses the leaves/some light misses the chloroplasts; Overlapping leaves/some leaves shaded; Other named factor may be limiting; (Some) trees are not in leaf all year round; Enzymes are not 100% efficient; *(max 4)*

Examiner's tip

'Suggest' always means that your good ideas will count. Stating the 'obvious' is usually beneficial. Here, it is clear that not all light actually hits the tree! Some is reflected, and some heats up objects. So dissipation as heat energy would be given credit.

(c) Heat/thermal;

Examiner's tip

Touch a leaf on a warm day. It is clear that some energy causes the leaf to warm up.

(4) **(a)** Protein/polypeptide;
(b) Rough endoplasmic reticulum has ribosomes; This is the site of protein synthesis;

Examiner's tip

The stem of the question gave you a cue. Rough endoplasmic reticulum has ribosomes and is involved in the assembly of polypeptides from amino acids.

(c)   Proteins/polypeptides move to the Golgi apparatus; Reference to protein modification; (Protein or polypeptides) enclosed in membranes to form vesicles; so most activity is in the vesicles after 45 minutes;   *(max 3)*

Examiner's tip

Follow the maximum radioactivity through each stage.
    RER → Golgi apparatus → Secretory vesicles
The sequence should remind you of protein synthesis.

(d)   Some amino acids are moving between sites; Amino acids are being broken down/metabolised; Proteins are used in other parts of cells; Proteins are also synthesised in the mitochondria;   *(max 3)*

Examiner's tip

If the radioactive amino acids did not equate to 100%, then this indicates that 5% were elsewhere in the cell. Did you know that they are broken down in respiration to release energy?

(e)   Exocytosis; Vesicles move to the cell membrane; Vesicle fuses with the cell membrane; contents are released outside of the cell;   *(max 3)*

Examiner's tip

Clearly if the radioactivity is lost to the cell, then secretion of a product made from radioactive amino acids has taken place.

Section B

(1)   *(Answers to part **(e)** of this question require continuous prose. Quality of written communication should be considered in crediting points in the mark scheme. In order to gain credit, answers must be expressed logically in clear scientific terms.)*

(a)   Blood grouping relies on the presence of agglutinogens/antigens;
on the cell surface membrane/plasma membrane of the red blood cells;
DNA is found in the nuclei; Only white blood cells have nuclei;   *(max 3)*

Examiner's tip

*You needed to recall that red cells are enucleate (they do not have a nucleus). The early technique relied on the **antigen** on the cell surface membrane to determine blood group. White blood cells have nuclei which can be used in more modern and accurate techniques.*

(b) (i) AB;   **(ii)** This showed that the blood has both antigen A and antigen B;

Examiner's tip

The presence of the anti-A antibody results in agglutination of red blood cells which have antigen A in their cell surface membrane. Agglutination is the clumping together of red blood cells. In the presence of antibody B the blood also agglutinates. The blood group must be AB!

(c) DNA of which the function is not known/does not code for a protein;

(d) (i) Restriction endonuclease/restriction enzyme;
(ii) Restriction enzyme cuts at specific base sequences; isolates non-coding DNA; sequence of bases repeated different numbers of times; number of bases in sequence determines the length of the piece of DNA; *(max 3)*

(e) Chains of DNA are separated; by heating; probe consists of complementary sequence; of DNA bases; radioactive; can be located as produces a shadow on photographic plate; *(max 5)*

(2) (a) Translation is the conversion of the base sequence on mRNA into the amino acid sequence of protein; It is carried out in the ribosomes; Each ribosome has a small sub-unit and a large sub-unit; The small unit has a region which will attach to codons of mRNA; A codon is a sequence of three bases in mRNA coding for one amino acid; An amino acid is activated; and is attached to a specific tRNA molecule; This carries the amino acid at one end and a specific anticodon at the other; The tRNA fits into the large sub-unit; There are two adjacent sites in each of the sub-units; As the ribosome passes along the mRNA, one codon at a time, tRNA with the appropriate anticodon fills the vacant slot; The amino acid forms a peptide bond; with the developing peptide chain held by the tRNA in the adjacent slot This continues until a stop codon is reached; *(max 10)*

(b) Each chromosome consists of two chromatids; Homologous chromosomes consist of one from each parent; and these pair up; The chromatids entwine forming chiasmata; at which point DNA may break and interchange/cross over; In animal cells, centrioles move to opposite poles; Spindle fibres form; In metaphase bivalents become attached to equator; by centromeres; Anaphase – chromosomes of each bivalent pulled to opposite poles; New spindles form at right angles to old; Centromeres divide (kinetochores) and attach to new spindles; Separated chromatids (now chromosomes) are pulled apart; Spindle and chromosomes disappear, nuclear envelope reappears; *(max 10)*

Mock Exam

Time: 1 hour 20 minutes   Maximum marks: 70

Section A

(1) The graph shows the rate of glucose absorption in and excretion from a mammalian kidney in relation to the glucose concentration in the plasma.

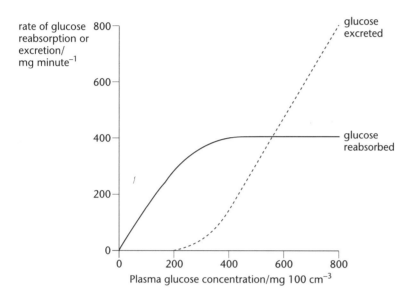

(a) Draw a line on the graph to show the rate of filtration of glucose in the renal capsule. **[1]**

(b) In which part of the nephron is glucose reabsorbed?

....**[1]**

(c) Explain the shape of the glucose reabsorption curve.

....

....

....

....

....

.... **[3]**

[5 marks]

[AQA A Specimen]

(2) The diagram below summarises the biochemical pathways involved in photosynthesis.

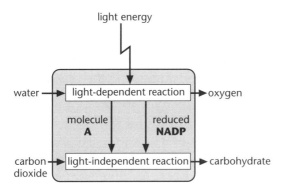

(a) Name Molecule **A** .... **[1]**

(b) **(i)** Describe how NADP is reduced in the light-dependent reaction.

....

....

....

.... **[2]**

(ii) Describe the part played by reduced NADP in the light-dependent reaction.

....

....

....

.... **[2]**

[5 marks]

[AQA B Specimen]

(3) The diagram shows the sequence of events which take place when the nucleus of a sperm enters the cytoplasm of an egg.

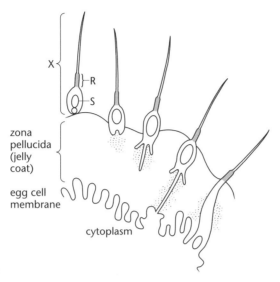

(a) Name the part of the reproductive tract in which these events take place.

.... **[1]**

(b) Complete the table to show the **names** of **R** and **S** in the diagram, a **structure** found in each and the **function** of these structures.

	Name	Structure	Function
R			
S			

[6]

(c) State **two** similarities between the process visible in the diagram and the process by which the male nucleus enters a plant ovule.

1. ....

....

2. ....

.... **[2]**

[9 marks]

[WJEC Specimen]

(4) The diagram shows a method which has recently been used to clone sheep.

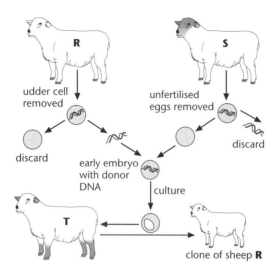

(a) Scientists used ewes from different pure breeding varieties in order to check that the procedure was successful at each stage and that the lamb produced was a clone of **R.**
Suggest what could have been deduced about the procedure if the lamb had been born:
   (i) with a black face,

   ....

   .... **[1]**

   (ii) with black legs.

   ....

   .... **[1]**

(b) **(i)** Suggest **one** reason why scientists did not think that it would be possible to clone sheep from udder cells.

   ....

   .... **[1]**

   (ii) Some scientists do not regard sheep produced in this way as a pure clone. Suggest **one** reason for this.

   ....

   .... **[1]**

(c) Suggest:

    (i) one concern in using this technology for human material.

....

.... **[1]**

    (ii) one possible benefit.

....

.... **[1]**

[6 marks]

[WJEC Specimen]

(5) Read the following passage.

Arrow poisons were widely used for hunting in South America where most of the powerful poisons such as curare originated. Charles Waterton carried out experiments on curare in the 1920s. He gave curare to a donkey which appeared to die ten minutes later.

5   The animal was then revived by artificial ventilation of its lungs with a pair of bellows and went on to make a full recovery. The experiment showed that injection of arrow poison into the blood stream causes death by respiratory failure. It is now known that curare competes with acetylcholine molecules for the receptors at neuromuscular junctions.

10 Since Waterton's work, many other chemicals have been discovered which also affect the nervous system, two of which are anatoxin and saxotoxin. Anatoxin also affects synapses. Its molecules are very similar to those of acetylcholine but they are not broken down by the enzyme, acetylcholinesterase. Saxotoxin is quite different and blocks the sodium channel proteins in nerve axons.

(a) During synaptic transmission, acetylcholine is released from the presynaptic neurone into the synaptic cleft.

    (i) Describe how an action potential brings about the release of acetylchloine into the synaptic cleft.

....

....

....

.... **[2]**

(ii) Describe how this acetylcholine may cause a new action potential to develop in the post synaptic nerve cell.

....

....

....

.... [3]

(b) Explain how injection of arrow poison into the blood stream causes death by respiratory failure (lines 6–7).

....

....

....

....

.... [3]

(c) **(i)** Explain how acetylcholinesterase is important in the functioning of a synapse.

....

....

....

.... [2]

(ii) Suggest an explanation for the fact that one of the symptoms of anatoxin poisoning is the excessive production of tears.

....

....

....

....

.... [3]

(d) Explain how saxotoxin may cause paralysis.

....

....

....

.... **[2]**

[15 marks]

[AQA B Specimen]

Section B

(1) (a) Microorganisms present in a rabbit's gut are able to digest carbohydrates in the plant material that they eat. The figure shows the biochemical pathways by which cellulose and starch are digested in the gut of a rabbit.

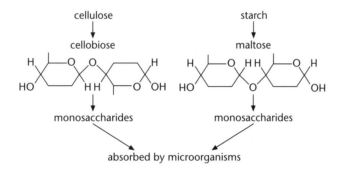

(i) Describe how a molecule of cellulose differs from a molecule of starch.

....

....

.... **[1]**

(ii) Draw a diagram to show the molecules produced by digestion of cellobiose.

[2]

(iii) Cellobiose and maltose are both disaccharides. Explain why amylase enzymes produced by the rabbit are unable to digest cellobiose.

....

....

....

....

.... [3]

(b) One way in which rabbits cause considerable damage to agricultural land is by competing for plant material that would normally be eaten by domestic animals. The table shows some features of the energy budgets of rabbits and cattle living under the same environmental conditions. All figures are kilojoules per day per kilogram of body mass.

	Rabbits	Cattle
Energy consumed in food	1272	424
Energy lost as heat	567	311
Energy gained in body mass	68	17

(i) What is the purpose of giving these figures per kilogram of body mass?

....

.... [1]

(ii) Explain the difference in the figures for the amount of energy lost as heat.

....

....

.... [2]

(iii) Use the information in the figure to explain why all the energy consumed in food cannot be converted to body mass or is lost as heat.

....

....

.... [2]

(c)   Rabbits were introduced to Australia in the middle of the nineteenth century. Their population grew rapidly and they are now major agricultural pests.

The table compares some features concerned with heat loss in cattle and rabbits at a temperature of 30 °C.

	Cattle	Rabbits
Percentage of body heat which is lost by evaporation	81.0	17.0
Core temperature of body °C	38.2	39.3

Use the information given in parts **(b)** and **(c)** of this question to explain:

(i) how evaporation helps cattle to maintain a constant body temperature;

....

....

....

.... **[2]**

(ii) the main way in which a rabbit would lose heat at an environmental temperature of 30 °C.

....

....

....

.... **[2]**

(iii) why rabbits are major agricultural pests in Australia;

....

....

....

....

.... **[2]**

(iv) why rabbits are better able to survive than cattle in the hot, dry conditions found in many parts of Australia.

....

....

....

....

.... **[3]**

[22 marks]

[AQA A Specimen]

(2) (a) The drawing below shows the larva of a species of caddis fly (*Hydropsyche* sp.) which is found in rivers.

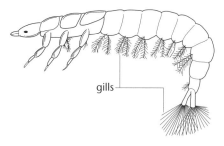

Note **two** distinct visible features of the gills which should facilitate gaseous exchange.

(i) ....

(ii) .... **[2]**

(b) The graphs below show the relationship between the rate of oxygen consumption and the temperature for the larvae of two species of caddis flies determined in the laboratory experiments.

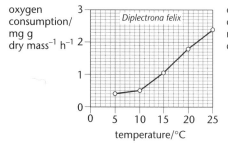

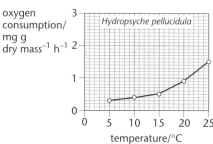

(i) Explain the effect of temperature on the rate of oxygen consumption of the caddis fly larvae (of either species).

....

....

....

.... [2]

(ii) What problem is posed to the caddis fly larvae as the temperature of the water increases from 5 °C to 25 °C?

....

....

....

.... [2]

(iii) Which species appears better adapted to living in water subject to temperature fluctuation? Explain your answer.

....

....

....

.... [2]

(iv) What type of pollutant might be expected to reduce the availability of oxygen in rivers? Explain your answer.

....

....

....

.... [2]

[10 marks]

[NICCEA Specimen]

Answers

How to use the mark scheme

Symbol	Meaning
;	A separate mark
/	An alternative answer acceptable
max	Mark scheme shows that marks available exceed the question value You can be awarded up to the maximum
<u>underline</u>	When the word or phrase is underlined then it must be given if a mark is to be awarded

Section A

(1) (a) The line continues from glucose reabsorption/the rate is directly proportional to concentration;

> **Examiner's tip**
>
> *Remember that in a healthy person **all** glucose which passes through the renal (Bowman's) capsule is reabsorbed in the proximal tubule. The mark scheme means that the line should be horizontal, continuing **back** along the glucose reabsorption line. The rate of glucose passing through the capsule = the rate of reabsorption.*

(b) First/proximal convoluted tubule;

> **Examiner's tip**
>
> *Remember here to give '**convoluted**'. Detail is the key.*

(c) Glucose is taken up by active transport; the more glucose there is, the more sites on the transporter proteins are in use; these are eventually saturated;

> **Examiner's tip**
>
> *The cells of the proximal tubule have transporter protein molecules in the cell surface membranes. Once they are all 'employed' taking glucose across, then the rate has reached maximum.*

(2) (a) Adenosine triphosphate/ATP;

(b) (i) Electrons are raised to a higher level; passed along a chain of hydrogen acceptors; uses protons or hydrogen ions; from photolysis/from water;   (*max 3*)

(ii) Reduces; GP/glycerate phosphate;

> **Examiner's tip**
>
> *The diagram is designed to stimulate your memory. The cues show the links between the light dependent and light independent reactions. In part (ii) the key fact is that reduced NADP gives up H to GP becoming, itself, oxidised back to NADP.*

(3) (a) Fallopian tube/oviduct;

(b)

	Name	Structure	Function
R	Mid-piece;	mitochondria;	ATP/energy for movement;
S	Nucleus;	chromosomes;	carries paternal genome/ paternal genes;

Examiner's tip

*This part of the mark scheme again shows the importance of giving detail. Did you give the correct function of the mitochondria. You may have given energy release, but needed to qualify this by **for movement**. Similarly the function of chromosomes carrying genes is not enough. **Paternal** chromosomes was required.*

(c) Both form a tube/both grow a tube; Nucleus travels along a tube into the egg; <u>Enzymes</u> are produced which allow a tube to grow; Only male nucleus enters egg/exine and rest of sperm remain outside; *(max 2)*

Examiner's tip

Both observation skills and memory were needed here. The diagrams show a process of which you may not be expected to know all of the detail shown. If you had revised fertilisation of a flower then you could identify the similarities. Without revision you would not achieve success!

(4) (a) (i) DNA was not removed from the egg of S/the DNA came from S;
(ii) Implantation of the manipulated embryo did not take place/T was already pregnant;

Examiner's tip

In this question it is necessary to interpret given information. The diagram shows that the DNA of S is meant to be discarded during the process.

(b) (i) Other studies had used early embryonic cells/other studies had used undifferentiated cells/udder cells are mature or differentiated/udder cells were thought to be inflexible;
(ii) The cytoplasm and DNA are from different individuals/ mitochondrial DNA is present from the enucleated cell;

Examiner's tip

*The key to part **(i)** is that once cells are differentiated then they were thought to have an established role. In answer to part **(ii)** you would need to know that there is a source of DNA other than from the nucleus. A more obvious answer is than the cytoplasm and DNA are from **two** organisms.*

(c) (i) The ethics of cloning individuals purely as a source of organs for transplants/the ethics of producing brain dead clones as organ banks/the ethics of using aborted fetuses for organ production/the responsibility of the welfare of cloned children/ the responsibility of the welfare of cloned animals/transmission of disease from animal to human and then to the population;

(ii) Production of perfectly matched tissue/avoidance of inherited diseases;

Examiner's tip

*The answer to **(i)** involves thoughts of ethics and consequences of using the technique. This is a good example of a question which needs careful consideration. Background reading and awareness of issues in the media would help you here.*

(5) (a) (i) Calcium ions enter the presynaptic membrane presynaptic neurone;
Vesicles fuse with the membrane and release contents;

(ii) (Acetylcholine) diffuses across the cleft; (Acetylcholine) fits the receptor sites on the membrane receptors; This causes the opening of sodium channels; Sodium ions enter the post synaptic neurones   *(max 3)*

Examiner's tip

*This question, so far, has depended on your recall of transmission across a synapse. The Examination Groups do not expect the exact words to statement in their scheme but you should convey the same meaning, e.g. for the second point you write that the acetylcholine **binds with** the receptor sites. This would be acceptable! Additionally a new action potential would be generated if enough Na^+ ions passed through the ion channels and reached a threshold value. Again this would be credited.*

(b) Binds with receptor molecules at neuromuscular junctions; acetylcholine molecules will not fit; Does not trigger muscular contraction; so that breathing is not possible;
                                                                              (max 3)

Examiner's tip

If the acetylcholine molecules cannot enter the receptor sites then transmission across the synapse is halted.

(c) (i) (Acetylcholinesterase) removes the acetylcholine;
so it will stop further action potentials/further impulses;

(ii) Tear production is controlled by the parasympathetic nerves; which have cholinergic synapses; anatoxin remains in the acetylcholine receptor sites and is not broken down; so it continues to send impulses;

Examiner's tip

Acetylcholinesterase is the enzyme which removes acetylcholine from the receptor sites. If it remained then the Na^+ channels would remain open and impulses will continue.

(d) Sodium ions cannot enter the neurone;
so no action potential can be sent/so no nerve impulse sent;

Examiner's tip

*No action potential could be generated because **enough** sodium ions must pass into the next neurone.*

Section B

(1) (a) (i) Cellulose is made from β-glucose and starch from α-glucose;
(ii) Diagram showing two recognisable monosaccharides;
Correct groups shown as a result of hydrolysis;

(iii) Cellobiose and maltose both have different shapes;
cellobiose will not fit/bind to/form enzyme-substrate complex with; active site of
the amylase;

(b) (i) To take account of the different sizes of the animals;
(ii) Heat is lost from the surface; the rabbit is smaller and has a greater surface area to
volume ratio;
(iii) Some of the energy consumed in food is used by the microorganisms;
For respiration (of microorganisms)/for the growth of microorganisms;

(c) (i) Evaporation requires (latent) heat; to change water into a vapour; this heat is drawn
from the animal's body; *(max 2)*
(ii) Must lose heat directly to a cooler environment by radiation/by convection;

(iii) Rabbits are not kept in check by natural predators/no predator of rabbits in
Australia; they consume more food per kg than cattle; competition for limited
resources; *(max 2)*

(iv) Rabbits have a higher body temperature than cattle; so do not need to lose as much heat; can lose more heat to the environment by radiation; do not rely as much on sweating as cattle; less dependent on drinking water; *(max 3)*

Examiner's tip

Data were given in the table. It was important to use the information on body temperature.
The fact that rabbits are more voracious herbivores means that they have a supply of water in the plant material consumed. More heat lost by radiation which does not require water loss!

(2) (a) (i) Large numbers/large surface area/branching;
(ii) External (and so are exposed to the aqueous medium);

Examiner's tip

Analysing the diagram it clearly shows the gills of high surface area. They are even labelled to give you further help.

(b) (i) Higher temperature increases the rate of respiration/increases the rate of enzyme action/increases the rate of metabolism; so the demand for oxygen increases;

Examiner's tip

Analysis of the graphs shows that as the temperature increases then so does oxygen requirement.

(ii) An increased water temperature would result in excessive demand for oxygen/increased water temperature reduces the amount of oxygen in the water (since there is a decrease in the solubility of gases in solution); so that the oxygen supply does not satisfy demand/is not sufficient;

Examiner's tip

Clearly as temperature increases there is less oxygen available, per unit volume, in the water medium. Additionally there is a greater need for more oxygen as respiratory rate increases.

(iii) *Hydropsyche pellucidula;* It has a lower oxygen demand at higher temperatures;

Examiner's tip

The graphs clearly display the different properties of the two species. The two species have similar oxygen demands at lower temperatures but at higher temperatures H. pellucidula *does not need as much oxygen.*

(iv) Hot water pollution from power stations;
which reduces the solubility of oxygen in water; *OR*
sewage (or other organic material) pollution;
oxygen removed by decomposing bacteria; *OR*
particulate matter;
photosynthesis reduced due to light blockage *OR*
fertiliser run off; decomposition of the resulting algal bloom;

For your notes

For your notes